Modern Polarographic Methods

Modern Polarographic Methods

Helmut Schmidt

Mark von Stackelberg

Institute for Physical Chemistry, University of Bonn, Bonn

Translated by

R. E. W. Maddison

1963

 Academic Press · New York · London

Translated from the German

Die neuartigen polarographischen Methoden—ihr Prinzip und ihre Möglichkeiten

published in 1962 by Verlag Chemie GmbH
as Monograph Number 77
to *Angewandte Chemie* and *Chemie-Ingenieur-Technik*

COPYRIGHT © 1963, BY ACADEMIC PRESS INC.
ALL RIGHTS RESERVED.
NO PART OF THIS BOOK MAY BE REPRODUCED IN ANY FORM,
BY PHOTOSTAT, MICROFILM, OR ANY OTHER MEANS, WITHOUT
WRITTEN PERMISSION FROM THE PUBLISHERS.

ACADEMIC PRESS INC.
111 Fifth Avenue, New York 3, New York

United Kingdom Edition published by
ACADEMIC PRESS INC. (LONDON) LTD.
Berkeley Square House, London W.1

LIBRARY OF CONGRESS CATALOG CARD NUMBER: 63-16983

PRINTED IN THE UNITED STATES OF AMERICA

Contents

Introduction 1

I. Methods with Controlled Voltage

   A. Stationary (Quasi-Stationary) Methods 5
      1. Conventional Direct-Current Polarography 5
      2. Differential Polarography 6
      3. Derivative Polarography 8
      4. Strobe Polarography (*Tastpolarographie*) 13
   B. Nonstationary Methods 16
      5. Oscillographic Polarography with Controlled Potential 16
        a Single-Sweep Method 17
        b Multi-Sweep Method 22
        c Advantages and Disadvantages of Oscillographic Polarography 24
      6. Polarography with Superimposed Alternating Voltage 29
        a Conventional Alternating-Current Polarography 29
        b Square-Wave Polarography 51
        c Alternating-Current Bridge Polarography 58
      7. Pulse Polarography 63

II. Methods with Controlled Current

      1. Oscillographic Polarography According to Heyrovský and Forejt 71
      2. Radio-Frequency Polarography 79

III. Combinations of the Above Methods 91

References 93
Author Index 97
Subject Index 99

Modern Polarographic Methods

Introduction

The polarographic method developed about 40 years ago by Heyrovský is still used with undiminished success; though it is unchanged in principle, numerous modifications of the classical method, some trivial, others important, have been published in recent years. It seems, therefore, timely to discuss these new procedures, to point out their advantages and disadvantages with respect to established polarographic methods, and, insofar as it is at present possible, to outline their significance for analysis and the solving of scientific problems.

The improvements striven for in these new methods have essentially the following intent:

1. *Increase in Sensitivity.* Although in recent years there has been a considerable advance in the development of apparatus, nevertheless the possibilities of modern amplifier and recording techniques have not by any means been fully utilized. The reason for this is that any further increase in the sensitivity of apparatus is pointless as long as over-all sensitivity is limited by the reaction which is being investigated. As is well known, this limitation of sensitivity is controlled in polarography by the capacitance current. For the build-up of the electrical double layer at a constantly maintained voltage, the increasing surface of the growing drop requires a transfer of charge, which is greater the further conditions are removed from the zero potential of the dropping-mercury electrode. (With an applied alternating-current voltage there is in addition an alternating-current charging current.) If the faradic current is of the same order of magnitude as the capacitance current, or is less, then its exact determination is impossible. Attempts to compensate the capacitance current were made from quite an early date, and from them a necessary improvement, particularly for analytical purposes, was achieved.[1] As, however, the

capacitance, and hence the charging current, are influenced not only by the continuously changing electrical double layer of the drop but also by adsorption and desorption of surface-active substances, and as the charging of the surface furthermore varies nonlinearly with the voltage, such an empirical compensation must always remain an expedient.

Consequently all improvements dealing with sensitivity must stem from elimination or compensation of the capacitance current. Only when this has been achieved, as for example with square-wave polarography, do other factors limiting the sensitivity acquire significance.

2. *Increase of Resolvability.* One great advantage of polarography compared with other analytical methods is the possibility of simultaneously determining two or more depolarizers independently of each other. If the half-wave potentials of two such substances lie very close together, then both waves merge, and an evaluation of the individual wave heights is no longer possible. As a measure of *resolvability* one can take directly the difference between the half-wave potentials of two depolarizers which can be resolved qualitatively or determined quantitatively with sufficient accuracy. This definition, however, is not adequate, because the resolvability is furthermore dependent on the ratio of concentration of both substances. If, together with one depolarizer A, there is present in large excess another one B, which is reduced at a more negative voltage, then the latter may be easily determined, as the diffusion current of A can be used as the reference current for B, and recorded at a correspondingly lower level of sensitivity. On the other hand, if A is present in large excess, it ought to be possible, after compensating the diffusion current of A, and with increased sensitivity, to determine B in this case too. In practice, however, it is not possible, because the second wave can no longer be separated from A in consequence of the increased sensitivity, which produces greatly magnified serrations in the curve corresponding to the fall of each drop. We call the ability to determine such substances in the presence of each other when the ratio of concentration is high the *separability*, to distinguish from resolvability as defined above.

3. *Special Applications; Miscellaneous.* Other improvements and potentialities offered by modern methods as opposed to classical polarography will be considered. For example, we have the applicability to kinetic investigations, or the study of adsorption phenomena. In addition, matters particularly important for analytical work, such

Introduction

as speed of operation, reproducibility, apparatus requirements, and so on, will be discussed.

It may seem surprising that the *accuracy* of individual methods is not specifically treated. The reason for this is that accuracy is closely linked with sensitivity and resolvability. Furthermore, accuracy is greatly dependent on the experimental conditions and the mechanical as well as electrical precision of the recording instruments employed. Such matters will be dealt with only very briefly in this monograph, which is essentially concerned with problems of method, and touches only the fringe of problems connected with apparatus.* In dealing with individual methods, any details relating to accuracy which have been reported in the literature will naturally be mentioned.

The following is a summary of the methods to be discussed. First of all a decision must be made as to what is to be understood by the concept "polarography." Delahay[3] defines it as a special case of voltametry in which a dropping-mercury electrode is used; Heyrovský[4] takes a wider view and regards it in general as the study of reactions at the dropping-mercury electrode. As to the convenience of these definitions, of which no notice is usually taken in the literature, no discussion will be made here. Potentiometric, conductometric, and high-frequency titration methods, as well as electro-gravimetric and electro-analytic potentiostatic separations, are not included in this work as they do not properly belong to polarography. Nor are coulometric methods and amperometric titrations considered. What remains may be conveniently summarized thus:

I. Methods with controlled potential
    A. Stationary (quasi-stationary) methods**
        1. Conventional direct-current polarography
        2. Differential polarography
        3. Derivative polarography
        4. Strobe polarography (*Tastpolarographie*)
    B. Nonstationary methods
        5. Oscillographic polarography
            a. Single-sweep methods
            b. Multi-sweep methods
        6. Polarography with superimposed alternating-current voltage
            a. Conventional alternating-current polarography

* A survey of developments concerning apparatus has been made by H. Schmidt.[2]
** Strictly speaking, direct-current polarography is not a stationary method; a quasi-stationary state is reached as a result of drops forming at the electrode.

        b. Square-wave polarography
        c. Alternating-current bridge polarography
    7. Pulse polarography
II. Methods with controlled current
    1. Oscillographic polarography according to Heyrovský and Forejt
    2. Radio-frequency polarography
III. Combinations of the above methods

I

Methods with Controlled Voltage

A. Stationary (Quasi-Stationary) Methods

1. Conventional Direct-Current Polarography

A knowledge of the principles, methods of working, and the applications of conventional direct-current polarography are assumed. Some data are given only to permit a comparison with the other methods about to be discussed.

The limit of sensitivity is reached in $10^{-5} N$ solutions, with an accuracy of 1 to 3 % in favorable cases.

Two waves may still be resolved qualitatively if their half-wave potentials are separated by at least 80 to 100 mV. For a quantitative determination the separation must be at least 100 to 120 mV.

At a concentration ratio of 10:1 the separability is already inadequate; at 50:1 and over the second substance can hardly be detected qualitatively.

These figures are naturally average values, because they are controlled not only by the method itself but also to a considerable extent by the type of apparatus employed and the general working conditions; they are dependent too upon the nature of the substances to be determined and the composition of the supporting electrolyte.

As direct-current polarography is only briefly treated in this book, attention is directed to the publications of H. W. Nürnberg and M. von Stackelberg[114] and of H. W. Nürnberg,[115] in which the subject is treated in greater detail.

2. Differential Polarography

In this method, developed by Semerano and Riccoboni,[5] two cells are arranged in opposition in a bridge, and the *difference* of the currents through both cells is measured (Fig. 1). If one of these cells contains,

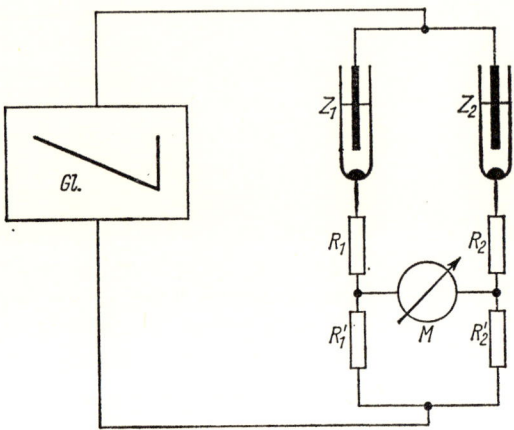

FIG. 1. Schematic diagram for differential polarography. Gl = Source of D.C. voltage; Z = cells; R = adjustable resistances; M = measuring instrument.

for example, only the supporting electrolyte, and the other has a depolarizer in addition, then the measuring instrument will register only the current produced by the deposition of the depolarizer if the electrical and mechanical dimensions of both cells are identical. The advantages of this method of working are:

1. The *sensitivity* is greatly increased, because not only is the capacitance current compensated, but so also generally is the whole residual current, including that resulting from conducting salts present as impurities.

2. The *separability* is also improved, especially if to the second cell with the blank solution not only a conducting salt but also a corresponding amount of the interfering ions is added. With exact synchronization of the two dropping-mercury electrodes the serrations caused by the falling drops are completely eliminated from the wave; this improves the separability.

3. In addition the *resolvability* is better, because a smooth polaro-

A. Stationary (Quasi-Stationary) Methods

gram without serrations makes it easier to recognize the change from one wave to the other, as compared with a polarogram which is modulated by the dropping process.

If in spite of all this there is little application of differential polarography, the reason lies in the difficulty of maintaining complete synchronism in the fall of drops from two capillaries.* Various ways have been tried to bring about complete synchronism by mechanical or electrical control of the drop, by streaming-mercury electrodes, or electrodes yielding drops extremely slowly. In spite of indisputable successes (Airey and Smales,[6] for example, were able to determine Zn perfectly in the presence of a 100-fold excess of Cd) any wide application for routine analytical determinations founders as a rule on the bad reproducibility of uniformly dropping twin electrodes.

Kelley and Miller[7] attempted to get around this difficulty by measuring the total current and residual current in sequence instead of simultaneously. The procedure is thus: First of all a polarogram of the residual current is taken. During the recording of the total current of the unknown depolarizer solution, the residual-current curve is tapped off synchronously by means of a photocell, and after conversion and amplification is subtracted from the total current, so that a polarogram is obtained directly which is free from the capacitance current and other contributions to the current made by the supporting electrolyte. By using this "curve follower" method in conjunction with the usual linear capacitance-current compensation, it was possible to determine lead in concentrations of 2×10^{-6} to $2 \times 10^{-7}\ M$ with an error of $\pm 20\ \%$. In any case the apparatus requirements are very considerable, and can only be justified when a very large number of individual determinations with the same supporting electrolyte has to be made.

A very elegant method, which eliminates only the capacitance current, has been developed by Barker.[8] He applies a small alternating-current voltage to the cell, so that the alternating current which flows is proportional to the capacitance of the electrical double layer. This current is amplified, rectified, and electronically integrated. The voltage derived therefrom gives the corresponding current needed for compensation in the measuring circuit proper. This method, too, requires a considerable amount of apparatus, but does, however, provide a considerable gain in sensitivity.

* Better results are obtained with the streaming-mercury electrode.

3. Derivative Polarography

In this method, as the name indicates, the differential quotient di/dE (or $\Delta i/\Delta E$) is plotted against E, instead of i against E; that is to say, instead of the usual polarographic curve, one records its *gradient*.*
Basically there are two possibilities here: Either the desired relationship $\Delta i/\Delta E$ is obtained and recorded directly by a suitable circuit arrangement, or the usual i/E relationship is first measured and then electrically differentiated. Heyrovský[9] used the former method, the principle of which is shown in Fig. 2. As in the case of differential

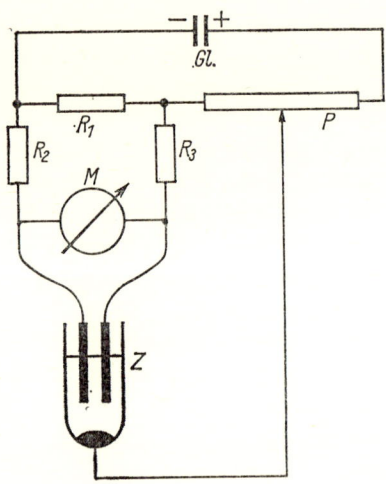

FIG. 2. Heyrovský's circuit for derivative polarography.

polarography, two synchronous dropping-mercury electrodes are used, but both dip into the same solution. The ratio of the resistances R_1 and R_2 is so chosen that the two capillary electrodes have a potential difference (ΔE) of about 10 mV with respect to the opposing mercury pool. The galvanometer M measures the corresponding difference of

---

* In the literature from English-speaking countries the expression "derivative" polarography is often used also for alternating-current methods. Even if the general form of an alternating-current polarogram does coincide to a considerable extent with the derived curve from a normal direct-current polarogram, there is nevertheless a sharp distinction between the two concepts, because they relate to two *fundamentally different* processes (see pp. 31, 32).

A. Stationary (Quasi-Stationary) Methods

the partial currents in R_2 and R_3 (Δi), so that the desired relation $\Delta i/\Delta E$ with respect to E is obtained directly. Instead of an ordinary galvanometer in such a bridge circuit it is possible to use a differential galvanometer, the two coils of which then replace the resistances R_2, and R_3 (Barendrecht[10]). The derivative polarogram so obtained (see Fig. 5) naturally shows a maximum at the half-wave potential, which is independent of the total current i. The aim of Heyrovský, namely to increase the *separability*, is thereby achieved. At the same time the *resolvability* is increased, because two overlapping waves become apparent as two separated peaks, while in the ordinary polarogram no limiting current is indicated between them. Apart from certain disadvantages in the circuit (ΔE is dependent on the galvanometer resistance and hence on its sensitivity; and there is always a leakage current through M), which were later eliminated by Airey and Smales,[6] this method, too, comes to grief on account of the difficulty already mentioned in connection with differential polarography, namely the requirement that both capillaries provide drops in complete synchronism. A particularly troublesome phenomenon of *beat* occurs when the drop-times are *almost* equal.

For this reason the second method is generally used (Heyrovský,[11] Lévêque and Roth,[12] Vogel and Říha,[13] Říha[14]). The principle of electrical differentiation may be explained by reference to Fig. 3. If a voltage U_E is applied to the input side, then the voltage U_C across the capacitor is practically the same as U_E, assuming that $1/\omega C \gg R$. The current i flowing through the capacitor is proportional to the periodic change in voltage; thus $i = C \cdot dU_C/dt \approx C \cdot dU_E/dt$. Finally across R we have the voltage $U_A = i \cdot R \approx R \cdot C \cdot dU_E/dt$. That is to say, the output voltage U_A is proportional to the differential quotient of the input voltage. The validity of this relationship becomes worse as R increases with respect to $1/\omega C$. The complete polarographic differentiating circuit for the capacitor method is shown in Fig. 4 (C = differentiating capacitor of about 2000 to 5000 μF; $R_1 - R_2$ = Ayrton shunt; $R_1 + R_2$ about 1 to 2 KΩ). As can be seen from what has been deduced above, the galvanometer M indicates the change in voltage drop across R_1,

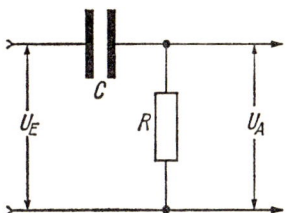

FIG. 3. Resistor-capacitor differentiating unit.

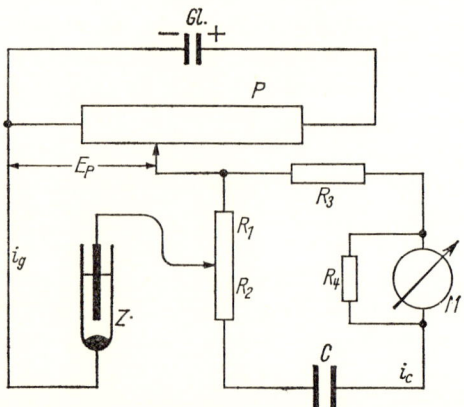

Fig. 4. Circuit for derivative polarography using the capacitor method.

i.e., $R_1 \cdot di_d/dt$, for which we can substitute $R_1 \cdot di_d/dE$, the desired relationship, if the voltage tapped off from P increases linearly with time.

Lingane and Williams[15] as well as others have dealt theoretically with the form and height of the derived polarogram. For a reversible electrode reaction* (for an irreversible reaction there is an additional specific factor) the maximum current flowing through M (i.e., the height of the peak) is given by

$$i_{C(\max)} = \left(\frac{\dfrac{nF}{RT} \cdot C \cdot R_1 \cdot i_d}{1 + \dfrac{nF}{RT} \cdot i_d \cdot R_K} \right) \cdot \frac{dE_P}{dt}$$

where i_d is the diffusion current and R_K the total resistance of the cell circuit, i.e., essentially the cell resistance $+ R_1$. This relationship is valid, however, only on the assumption that the time constant of the capacitor circuit $C \cdot (R_2 + R_3 + R_M)$ is small compared with dR_K/dt. Otherwise i_C lags behind di_g/dt, that is to say, the peak of the derivative polarogram lies negative with respect to the half-wave potential, the derivative curve becomes unsymmetrical, and the maximum is lower than is to be expected from theory. In addition, not only is the polarographic wave differentiated, but also each single drop, whereby unusually large kicks or serrations are produced. This necessitates

* The derivation of this and subsequent relationships cannot be given within the limits of this work. Reference must be made to the original articles.

A. Stationary (Quasi-Stationary) Methods

additional heavy damping of the detecting instrument, which is equivalent to increasing the response time of the recorder, whereby the distortion and lowering of the curve as described above are still further intensified. Jäckel[16] has given a detailed mathematical treatment of these distortions.

The above equation shows yet another important fact, namely, that the current maximum of the derivative curve is *not* proportional to the concentration. Proportionality occurs only when the second term in the denominator is negligibly small compared with unity, i.e., if both i_d and R_K are small. Hence we have the requirements of working with the lowest possible concentrations, and of keeping the cell resistance and other resistances in the circuit as small as possible. Each requirement is possible only within limits, because the sensitivity of derivative polarography is less than that of conventional polarography anyway. Finally let it be mentioned that the proportionality between $i_{C(max)}$ and the capacitance C indicated by the above equation is not exactly valid, as with an increase in the value of C the time constant $C \cdot (R_2 + R_3 + R_M)$ also increases; this causes a lowering of the maximum, as has already been set out above.

Experiments to remove these shortcomings have not been wanting. Mention may be made of a special filter circuit (*T-RC* filter) for damping the recording instrument, by means of which the serrations caused by the falling drops may be selectively damped by the use of an appropriately chosen time constant, without altering the shape of the heights of the polarograms (Kelley and Fisher[17]); the production of extremely short drop-times (down to 0.1 sec) with a capillary specially constructed to minimize the serrations produced by the falling drops (Skobets and Kavetskii[18]); or mechanical knocking-off of the drops (Airey and Smales,[6] and Wolf[19,20]); and finally the use of a transformer instead of the differentiating capacitor (Paulik and Proszt[21]). If there is a change in the current i_p flowing in the primary winding of a transformer, then the flux changes in accordance with the equation

$$\frac{d\Phi_p}{dt} = K_p \frac{di_p}{dt}$$

where K_p is a constant depending on the dimensions and magnetic properties of the winding. The E.M.F. induced in the secondary winding is

$$E_s = -K_s \cdot \frac{d\Phi_s}{dt}$$

In a transformer the flux in both windings is the same, so $\Phi_p = \Phi_s$, and

$$E_s = -K_s \cdot K_p \frac{di_p}{dt}$$

If R is the resistance of the secondary circuit—i.e., generally speaking, the resistance of the secondary winding and the galvanometer—then we can write

$$i_s = -K_p \cdot K_s \cdot \frac{1}{R} \cdot \frac{di_p}{dt}$$

and, with linear change of voltage, if all the constants are assembled together as K

$$i_s = -K \cdot \frac{di_p}{dE}$$

which is once again the required relationship.

The advantage of this method is that there is no distortion, and hence no lowering of the derivative curve, but a *symmetrical* peak. The reproducibility of the wave heights and the concentration proportionality is accurate to $\pm 2\%$ according to the authors. There is nevertheless an enlargement of the serrations caused by the falling drops, which must be eliminated by additional damping, though the consequence may be some wave distortion by the recorder. The electrical characteristics of the transformer primary must match those of the polarographic current circuit as closely as possible, and similarly those of the transformer secondary with respect to the recording instrument. Furthermore, because of the relatively small di/dE values, the transformer must have a very high inductance, a requirement that can lead to a component of considerable size.

From all the above, it follows that derivative polarography does indeed in principle offer advantages, but in practice is of little utility: the twin-electrode method, because of the difficulty, already discussed under differential polarography, of ensuring synchronism of the drops falling from the capillaries; and the capacitor method because of fundamental sources of error, which can be only partially eliminated by the transformer method, and then only with considerable expense. For exact quantitative measurements derivative polarography is definitely unsuitable, but it can give good service in resolving very closely adjacent waves, which even with a separation of 50 mV are clearly distinguishable as peaks. An example is shown in Fig. 5.

The *separability* is indeed likewise better, but only if the recorder is heavily damped, and a distortion of the peak is accepted into the

A. Stationary (Quasi-Stationary) Methods

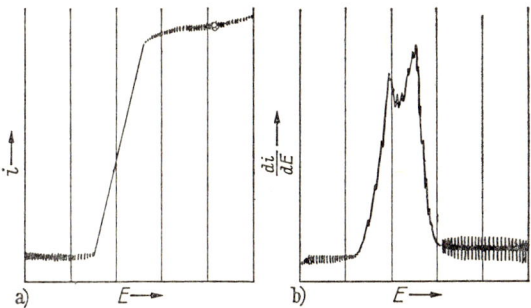

FIG. 5. (a) Conventional and (b) derivative polarogram of a solution of $2 \times 10^{-4}\,N$ InCl$_3$ and $5 \times 10^{-4}\,N$ CdCl$_2$ in $2\,N$ LiCl (Lévêque and Roth[12]).

bargain. The sensitivity, according to the method selected, is 1/10 to 1/100 that of direct-current polarography.

It may be mentioned that the principle of electrical differentiation is used not only in direct-current polarography, but finds frequent application also as an additional aid for increasing resolvability and separability in other methods yet to be discussed.

4. Strobe Polarography (*Tastpolarographie*)

Conventional polarography is carried out with a heavily damped recording instrument, which measures the average current $\bar{i}$ during the drop-time. The current i_d resulting from diffusion effects at the drop increases almost proportionally to $t^{1/6}$, where t is the age of the drop; the capacitance current i_C resulting from charging the electrical double layer is proportional to $t^{-1/3}$ (Fig. 6). This means that when the

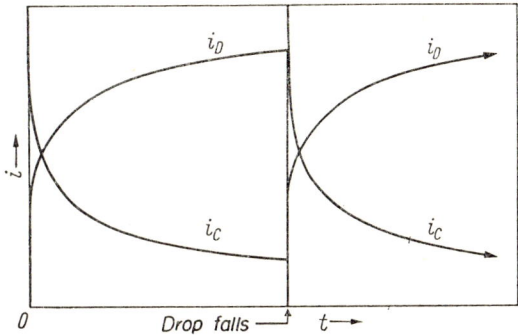

FIG. 6. Current-time curves for single drops. i_D = Diffusion current; i_C = capacitance current.

drop starts to develop the ratio i_d/i_c is considerably smaller than it is toward the end of the drop-time, when the increase in surface of the drop is small and the capacitance current is small. It is obvious, therefore, that the current should not be measured throughout the whole life of the drop, but only during a small fraction toward the end.

Wåhlin and Bresle[22] were the first to accomplish this by so controlling the drop-time and recorder with a mechanical interrupter that the current was measured and recorded only during a definite and constant fraction (about 1 sec) of the total drop-time, which was likewise kept constant (about 5 sec), while the recorder was out of operation during the remaining 4 sec. Barker and Jenkins[23] used the same principle in their square-wave polarograph (see below). Finally Kronenberger et al.[24] have developed an up-to-date polarograph working on this principle; and they have originated the term *"Tastpolarographie"* (strobe polarography). In contrast to the apparatus of Wåhlin, with the strobe polarograph* the synchronization of the fall of the drop and control of the recorder, paper feed, and potentiometer is effected

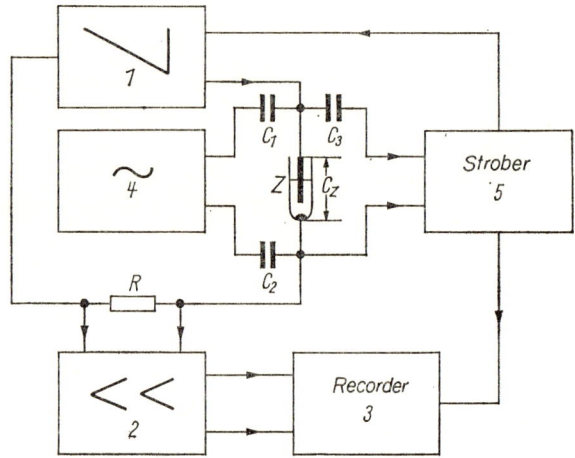

FIG. 7. Schematic diagram for strobe polarograph (*Tastpolarograph*).

electronically, so that by using suitable R-C components the start and finish of the interval which is being sampled can be varied within wide limits.

The operating principle can be explained by reference to Fig. 7. The

* Made by Atlas-Werke AG, Bremen, West Germany.

A. Stationary (Quasi-Stationary) Methods

polarographic current circuit consists as usual of the D.C. voltage source (1), the cell Z, and the external resistance R. The voltage across this, which is to be measured, is amplified by the amplifier (2), and registered by the recorder (3). In addition an A.C. voltage, which is so adjusted that it has practically no effect on the diffusion current, is applied to the cell through the D.C. blocking capacitors C_1 and C_2. When the drop falls the impedance suddenly changes. This impulse is coupled to the strober (*Tastgerät*) (5) from the mid-point of the voltage divider constituted by C_1 and C_Z (= capacitance of cell), and across C_3. A retardation circuit, having an adjustable time constant whereby the start of the interval to be recorded can be determined, is then actuated. A second retardation circuit, whose time constant controls the *duration* of the interval, finally operates a relay, which controls the advance of the potentiometer and the feed of the recording paper.

The advantages of strobe polarography are as follows:

1. The *sensitivity* is increased, as the ratio i_d/i_c is greater than in conventional polarography, and indeed is the greater, the longer the drop-time and the closer the interval which is being recorded lies to the drop point.

2. *Damping* of component apparatus is *unnecessary*, as even with long drop-times the oscillations are small. Polarograms by this method are simpler, and can be evaluated more exactly than in other forms of polarography, where excessive serrations on the wave as well as too much damping can falsify both the form and height of the polarographic curve (Strehlow[25]).

3. In contrast to conventional polarography, strobe polarography provides total current and residual current curves which decay horizontally. This means that the limiting current even in the extreme negative region is independent of the potential, and that all polarograms, which differ only with respect to concentration, are parallel in the region of the limiting current, whereby their evaluation is greatly facilitated and, above all, is more exact.

4. Because the capacitance current varies almost linearly with potential, the use of linear capacitance-current compensation is much more favorable, and consequently the utilizable sensitivity is still further increased. However, the increase in sensitivity mentioned under paragraphs 1 and 4 relates only to the elimination of the capacitance current; the other portions of the residual current (e.g., those parts of the faradic current resulting from impurities or traces of oxygen)

naturally are not compensated. Hence the resolvability and separability are improved only insofar as the waves have their natural gradient, because of the absence of damping in the apparatus, though the waves may be modified by small serrations and a linear falling-off of the diffusion current.

Strobe polarography cannot be used directly for kinetic investigations, because the established calculations of kinetic currents are based on the current flowing throughout the drop-time. In such cases the strober (*Tastmechanismus*) must be taken out of circuit, so that the recording may be made in the conventional manner.

Exact figures from comparative measurements regarding increase of sensitivity, resolvability, and separability are so far hardly available (Elbel[26]). Careful investigations by Bresle[27] on current-time curves provide information on the accuracy and reproducibility of strobe polarographic methods.

B. Nonstationary Methods

5. Oscillographic Polarography with Controlled Potential

The term "oscillographic polarography"* needs defining first of all in order to limit its meaning. The use of a cathode-ray oscillograph as a detecting or registering instrument is not in itself a new method. The oscillograph (cathode-ray or Braun tube) is very suitable as the registering instrument because of its freedom from inertia, its insensitivity to vibration, which is particularly disturbing when a mirror galvanometer is used, and its inherent high degree of sensitivity. There is only the one disadvantage that the rather small oscillograms must be preserved photographically, unless they are evaluated directly from a screen with persistent luminescence. For certain rapid reactions (e.g., *i-t* curves at single drops, or the exact representation of current maxima) ordinary galvanometers or recorders cannot be employed, and recourse must be had to the cathode-ray oscillograph, which is free from inertia.

The stimulus to develop true oscillographic polarography, as will be discussed here, gave rise to reaction-kinetic problems. If, in a solution

---

* Oscillographic polarography according to the method of Heyrovský and Forejt, which measures the potential-time relationship with constant current, has nothing to do with what is described here, and will be discussed in a later section.

B. Nonstationary Methods

under polarographic examination, there occur any chemical reactions which alter the concentration conditions within a few minutes or seconds, or even sooner, then ordinary polarography is no longer applicable as an analytical method for these participants in the reaction, because the current indicated represents the *mean* over a large number of drops, and consequently the time during which the polarizing voltage is applied must be large compared with the drop-time.

A decrease in the drop-time is possible only within very narrow limits. Matheson and Nichols,[28] who probably were the first to use oscillopolarographic methods, originated an arrangement using a drop-time of 1/30 sec, which already exhibits the main features of modern cathode-ray polarography. As the voltage sweep is synchronized with the drop-time, each drop gives directly a complete polarogram, which in its usual form appears as a stationary image on the screen of the oscillograph. This method is really very interesting, although it has acquired no practical significance. It represents a transition in that the drop-time and the time of voltage application are equal, whereas in classical polarography the time during which the polarization voltage is effective is large compared with the drop-time, and in oscillographic polarography it is small. In the second case very rapid changes in the solution subjected to polarographic examination may in fact be followed qualitatively and quantitatively, provided that, during the normal drop-time, the period of the voltage sweep is increased many hundreds of times, and complete polarograms are secured in fractions of a second.

It is clear that this technique produces completely new problems from the point of view of theory and apparatus. The particular polarographic method in which the duration of the voltage sweep is short compared with the drop-time will be discussed below. It is convenient to divide the matter into (a) single-sweep method and (b) multi-sweep method.

a. Single-Sweep Method

This method is of significance only for analytical purposes, and derives its name from the fact that each drop during its life receives only *one* (usually saw-toothed) impulse of polarization voltage. The production of this impulse of voltage increasing linearly with time is markedly different from that adopted in the technique of conventional polarography, where it is usual to employ motor-driven potentiometers giving a voltage sweep time of the order of 0.2 V/min. The production of very rapid voltage sweeps (for example, a frequency of 50 cycles with an amplitude of 2 V corresponds to 6000 V/min) necessitates electronic impulse generators, which permit really good linearity and variation of the sweep frequency within wide limits. The saw-toothed voltage must also have exact linearity, besides being as constant as possible in duration and amplitude. Whereas in the case of an ordinary polarogram the diffusion current is independent of the speed of the voltage sweep, in oscillographic polarography the value of the peak current is influenced by the speed of the voltage sweep, as will be shown later.

Finally, the choice of the moment at which the impulse is applied to the dropping-mercury electrode is of utmost importance. In normal polarography the current represents an average value over a large number of single drops, but in oscillographic polarography it is controlled by the surface area of the drop which increases at a rate proportional to $t^{2/3}$. If the duration of the impulse is very short, then the increase in surface area of the drop during the voltage sweep (ΔO) may be neglected, and the current will depend only on the instantaneous value of the surface area. This condition is realized, the shorter the sweep time τ, the longer the drop-time, and the nearer to the end of the drop-time that the impulse is applied. These conditions are shown schematically in Fig. 8. In order to obtain reproducible

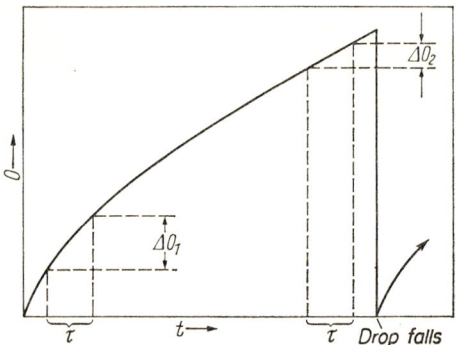

FIG. 8. Dependence of increase in surface area (ΔO) during time (τ) of voltage sweep on age of drop.

results, and this is naturally a prerequisite for all analytical applications, the moment at which the voltage is applied must always be the same; that is to say, the impulse and drop-time must be exactly *synchronized* either mechanically (rarely) or electrically (usually).

B. Nonstationary Methods

When one remembers that the current, before it is registered, must be amplified to a very considerable degree (and strictly linearly), and that all components necessitate a voltage supply which is carefully stabilized, screened, and freed from background, then one has some idea of the requirements in respect of apparatus for setting up a cathode-ray polarograph. Some figures will make this clear.[29] If the oscillogram is to have an accuracy of 1 %, then the level of interference must be under 2×10^{-4} V.

Apparatus and technical details will not be considered further here. Detailed information will be found in the writings of Cruse and Heberle,[29] as well as other authors.[30-35]

We will consider in more detail here the form and special features of oscillographic polarograms. On application of a rapidly increasing negative voltage to an electrode having for practical purposes a constant surface area, a current starts to flow when the reduction potential of the depolarizer is reached; and it becomes greater with increasing negative potential as in the case of an ordinary polarogram. This is dependent on the increasing concentration gradient at the surface of the electrode (curves 1, 2, and 3 of Fig. 9). However, the increasing

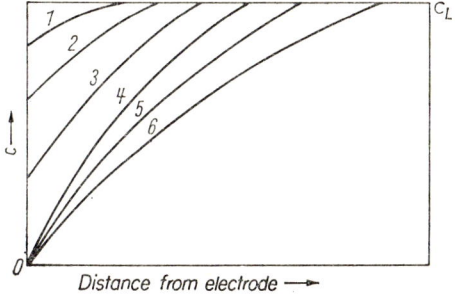

Fig. 9. Consecutive concentration curves of the depolarizer with continuously increasing voltage at an electrode with constant surface in a still depolarizer solution (schematic). Maximum current shortly before curve 4. C_L = Concentration of solution.

depletion of the depolarizer in the proximity of the electrode surface soon makes itself apparent. The diffusion layer increases; the concentration gradient (curves 4, 5, and 6) and the current become smaller. In conventional polarography, where the voltage increases very slowly, the depletion of the depolarizer is continually compensated by the detachment of the drop.

This does not take place with the rapidly increasing voltage impulse. When one peak value has been passed, the current falls asymptotically toward zero (Fig. 10). At some distance from the peak $i \sim t^{-1/2}$. The peak current is greater than the diffusion current of an ordinary

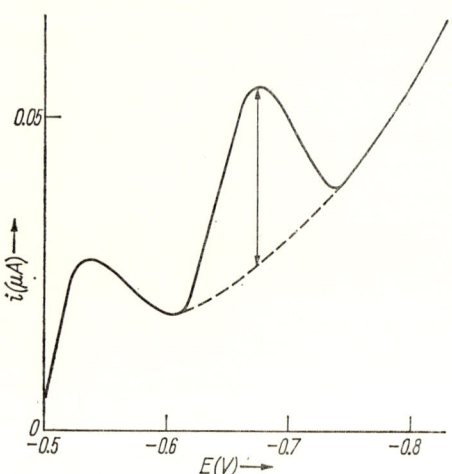

Fig. 10. Polarogram obtained by the single-sweep method. Solution: 0.18 mg/liter Cd^{2+} in 1 M HCl. (Ferrett et al.[44])

polarogram, because a still unconsumed store of depolarizer is suddenly utilized at an electrode surface which is already fully developed. The following relationship holds between the peak potential and the half-wave potential:

$$E_S = E_{1/2} - 1.1 \frac{RT}{nF}$$

The peak potential is therefore $28/n$ mV more negative than the half-wave potential. An equation for the peak current has been derived independently by Randles[30] and Ševčik[36] (cf. also Delahay[37]), the derivation of which will not be considered here. On purely qualitative grounds it is easy to recognize that the peak current must depend not only on the concentration c, the electrochemical equivalence n of the depolarizer, and the surface area of the electrode (proportional to $m^{2/3} \cdot t^{2/3}$), but also on the two factors which affect the concentration gradient, namely, the diffusion coefficient D and the speed of the voltage sweep $dE/dt = v$. In the Randles-Ševčik equation

$$i_S = k \cdot n^{3/2} \cdot m^{2/3} \cdot t^{2/3} \cdot D^{1/2} \cdot v^{1/2} \cdot c$$

B. Nonstationary Methods

the quantity v appears as a new essential factor together with the parameters of the Ilkovič equation (the constant k naturally has a different value). Of particular importance for analytical work is the fact that here, too, there is a linear relationship between peak current and concentration.

The Randles-Ševčik equation may be confirmed very well experimentally, but only under certain limiting conditions, a knowledge of which is indispensable for a sensible application of oscillographic methods. First of all, the electrode reaction must be *completely reversible*. Delahay[38] has concerned himself also with the case of irreversible reduction and has been able to show that here, too, the peak current can be given by the Randles-Ševčik equation, if one introduces the transfer coefficient α, and replaces the term $n^{3/2}$ by $(\alpha \cdot n_a)^{1/2}$, where n_a represents the number of electrons transferred in the stage determining the reaction. In the strictly reversible case $\alpha = 1$, and $n_a = n$; the more irreversible the reaction, the nearer α approaches zero, i.e., the peak current falls. As it is not possible to calculate α one is always forced to use calibration curves in practice, if one is not sure that the reaction is in fact strictly reversible. On the other hand, the Randles-Ševčik equation does offer a welcome possibility of testing the degree of reversibility of an electrode reaction. According to Cruse and Heberle[29] it is more convenient to observe the dependence of current on the square root of the speed of the voltage sweep, whereby deviations from linearity are shown up even with moderately irreversible reactions.

A second condition for the validity of the Randles-Ševčik relationship is that only *soluble products* participate in the electrode reaction. Berzins and Delahay[39] have also considered the case in which the reduction product is deposited at the electrode in the solid state. The agreement between experiment and theory is, however, unsatisfactory, because it is not possible to make any exact prediction about the change in activity with time of the precipitate at the surface of the electrode. As this activity is greatly dependent on the working conditions, the reproducibility is very bad, and for this reason the applicability of the method to analytical purposes in the case where there is deposition of solid reaction products is very restricted.

In this connection attention should be drawn to the fact that maxima-suppression agents, such as surface-active agents generally, are able to reduce the current peak very considerably,[32] and for this reason are to be excluded as far as possible from oscillopolarographic investigations. The cause of this phenomenon according to Strassner and Delahay[40] is to be ascribed to the fact that such substances influence the kinetics of the electrode reaction to such a degree that the total reaction becomes irreversible.

Finally, in using the Randles-Ševčik equation, it must be remembered that the peak currents are not measured against zero current, but against a reference current, that is to say, the capacitance current; and, in the case of mixed depolarizers, the current derived from previously reduced substances must be taken into account. These matters will be further considered in discussing sensitivity and resolvability.

b. *Multi-Sweep Method*

In the impulse method, as fully described above, the saw-toothed impulse is applied once only to each drop, and then as near as possible to the end of the drop-time, because the increase in surface area during the impulse is then least. Each drop therefore provides one polarogram only in its lifetime. The sweep voltage may also be applied permanently to the cell. Each drop will then receive *many* saw-toothed impulses, which will follow continuously one after the other. Each of these impulses provides a polarogram whose height is naturally a function of the electrode surface at that moment. As the saw-toothed voltage is applied to the cathode-ray tube as a time base, all these curves will be traced synchronously one above the other. The polarogram will appear on the screen as a whole host of curves of different height, the top one corresponding to the last impulse received during the life of the drop when its surface area is greatest. This method also has practical application,[28,29,31,32] and is called the *multi-sweep method*. Compared with the single-sweep method it has the advantage of greater simplicity from the point of view of apparatus, because the relatively difficult synchronization of the impulse with the growth of the drop is absent. To this advantage, however, must be added a number of disadvantages. First of all, everything that has been said about the degree of reversibility of the reaction and the solubility of the reaction products in connection with the single-sweep method naturally applies here too. In addition there are the following complications. If a calibration curve is taken by the multi-sweep method, that is to say, peak currents plotted against concentration, then a straight line is obtained, corresponding to a linear relationship. On comparison of this calibration curve with that obtained by the single-sweep method under otherwise exactly similar conditions, it is found that the straight

B. Nonstationary Methods

line obtained by the multi-sweep method has a smaller gradient, showing that all values of the peak current lie below those given by the single-sweep method. The Randles-Ševčik equation is *no longer* valid. An explanation of this phenomenon is to be found in the fact that the initial requirement that there should be no concentration fall at the moment of applying the impulse is no longer fulfilled. The voltage jump back to zero at the end of the impulse does indeed set up a reverse reaction, which tries to compensate the fall in concentration that has developed, and to an extent that is nearer to completion the more reversible the electrode reaction. If the individual impulses follow immediately one after the other, then this equalizing of the concentration is incomplete, and each fresh impulse finds a fall in concentration whose magnitude steadily increases with the number of impulses.

The existence of a similar *depletion effect* also in conventional polarography has been shown by Airey and Smales,[6] and Hans and Henne.[41]

In order to allow sufficient time for diffusion and the reverse reaction to equalize the concentration Delahay and Perkins[34] introduce a rest period between each impulse, during which the electrode is at zero potential. This is relatively easy to accomplish electrically with a sweep-voltage controller and appropriate adjustment of the zero level. In the case of a highly irreversible reduction even this procedure is of limited effectiveness.

A further difficulty lies in the fact that reproducible peak currents can only be expected when the sweep frequency is an integral multiple of the drop frequency; otherwise, the maximum value of the voltage of the last impulse does not coincide with the greatest surface area of the electrode. As the drop-time depends on the potential, and can be irregular for other reasons as well, its exact adjustment to the sweep frequency is usually imperfect. The consequence of *imperfect* synchronization is very troublesome surges.

If one wants to achieve useful results by the multi-sweep method, it is necessary (1) to allow sufficiently long pauses between the individual impulses, and (2) to synchronize the sweep frequency and drop-time exactly. The sole advantage of the multi-sweep method, namely, the need for a smaller amount of equipment, is thereby lost; and the single-sweep method is seen to be a special case of the multi-sweep method, where the rest period is practically equal to the drop-time, and the single impulse per drop is synchronized with the growth of each drop. For analytical purposes the single-sweep method alone is of interest.

Mention may be made also of the use of triangular impulses[36,38,42,43] instead of saw-toothed impulses.

In this case the voltage decreases at the same rate as it increases, and not suddenly as with the saw-toothed impulse. Two polarograms are thus produced, the second one corresponding to the oxidation of the substance reduced during the period of increasing voltage, and thus representing a mirror image of the reduction curve when the reaction is strictly reversible. Figure 11 shows an oscillogram obtained in this

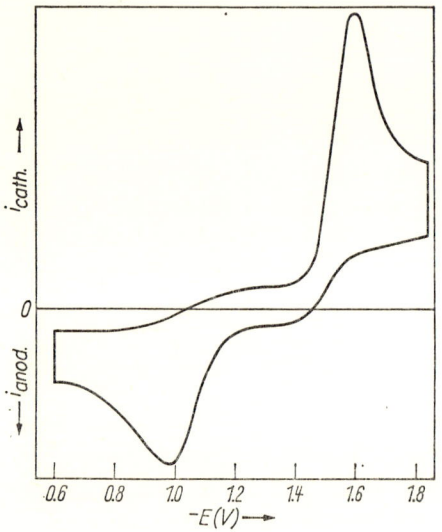

FIG. 11. Oscillogram obtained with triangular-shaped voltage impulses (Ševčik[36]). Solution: 10^{-3} M Zn^{2+} in 2 M NH_3/NH_4Cl—Buffer.

manner for the irreversible reduction of zinc, in which the great difference between the half-wave potentials corresponding to the reduction and oxidation stages may be seen.

c. *Advantages and Disadvantages of Oscillographic Polarography*

Now that the most important points concerning this method have been mentioned, it remains to consider the advantages and disadvantages of oscillographic polarography.

First of all, *sensitivity*: In polarography the current is determined by the concentration gradient at any moment. It decreases, however, proportionally with time while voltage is applied, because the diffusion

B. Nonstationary Methods

layer steadily increases in thickness. At the same time in conventional polarography the surface area of the drop increases, so that the maximum of the concentration gradient coincides with the minimum of the surface area of the electrode, and *vice versa*. The superimposing of both effects provides the well-known $t^{1/6}$ relationship between current and drop life. The current which is measured polarographically is a mean value taken over all the individual drops.

In oscillographic polarography the surface area of the drop in the single-sweep method may be regarded as constant. It is therefore the same for every concentration gradient; and, what is most important, it is the maximum, because each impulse is applied at the end of the life of the drop. The peak current is therefore considerably greater than the diffusion current of conventional polarography.

As was emphasized earlier, the sensitivity is determined not by the absolute value of the current, but by the ratio of the faradic current to the capacitance current. In general, it is given by

$$i_C = \frac{dQ}{dt} = \frac{d[C(E, t) \cdot E(t)]}{dt}$$

$$= \frac{dE}{dt} \cdot C + \frac{dC}{dt} \cdot E + \frac{dC}{dE} \cdot \frac{dE}{dt} \cdot E$$

$$= \frac{dC}{dt} \cdot E + v \cdot \left(C + \frac{dC}{dE} E \right)$$

where C is the capacitance of the electric double layer, and $dE/dt = v$ is the sweep velocity of the polarization voltage. In conventional polarography the second term may be neglected, as v is very small (zero in polarometric measurement). If dC/dt were constant, a linear relationship would exist between i_C and E. Unfortunately dC/dt is a function of E, so consequently any capacitance-current compensation increasing linearly with the polarizing voltage must be incomplete.

On the other hand, in oscillographic polarography the first term of the above equation may be neglected, as the surface hardly changes during the very rapid voltage sweep, and consequently dC/dt remains very small. The capacitance current is therefore given by the second term, between which and the rate of the voltage sweep there exists a linear relationship. On the other hand, we know from the Randles-Ševčik equation that the faradic peak current increases proportionally to the square root of speed of the voltage sweep. The ratio of the faradic current to the capacitance current is therefore not constant, but depends on v. These conditions are shown on Fig. 12 for two dif-

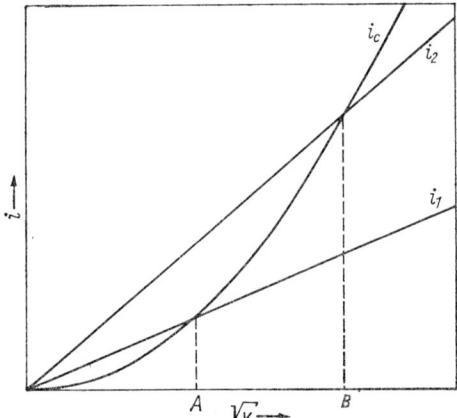

Fig. 12. Dependence of the faradic current and the capacitance current on the speed of the voltage sweep.

$i_1 = k \cdot c_1 \cdot v^{1/2}$ = faradic current at concentration c_1
$i_2 = k \cdot c_2 \cdot v^{1/2}$ = faradic current at concentration c_2
$i_C = k \cdot v$ = capacitance current

ferent concentrations. From this it can be seen that (1) the smaller the value of v, the greater is the ratio i_F/i_C (i.e., the effective sensitivity); and (2) the lower the concentration (A and B in Fig. 12), the sooner does the capacitance current attain the value of the faradic current. It is not the case, as one would suppose from the Randles-Ševčik equation, that the sensitivity can be increased at will by increasing the value of v; on the contrary, v must be kept as small as possible, and that to a greater extent the lower is the concentration of the solution.

From this it follows that compensation of the capacitance current is still much more difficult than in conventional polarography. On the other hand, the increase in the faradic current compared with normal polarography is so considerable that there still remains a very noticeable gain in sensitivity, in fact a good 10 times for an appreciable concentration. At much lower concentrations of depolarizer the disturbing effect of the capacitance current becomes apparent, in that an exact extrapolation of the reference current becomes difficult. Moreover the noise of the oscillograph has already reached the level of the current to be measured.

The *resolvability* and *separability* are better because (1) the polarograms are not modulated by the serrations resulting from the fall of

B. Nonstationary Methods

the drop; and (2) peak currents can be separated better than current waves, as was shown above in discussing derivative and differential polarography. This result can be further improved if the polarograms are differentiated electronically by means of suitable R-C units. In this way one can get down to 40 mV for resolvability; and to a concentration ratio of 100 to 400:1 for separability (Ferrett et al.[44]). These figures are, of course, valid only with certain reservations, when substances are to be determined quantitatively. Where several peak currents follow one after the other, each individual current curve forms the base line for the next one. If the individual peaks lie very close to each other (Fig. 13), they are easily recognizable as such, but their subsequent course is obscured by the rise to the next peak, and must therefore be extrapolated. The degree of uncertainty attaching to the results through this very inaccurate extrapolation is greatly accentuated by electronic differentiation, as was pointed out in the section dealing with derivative polarography.

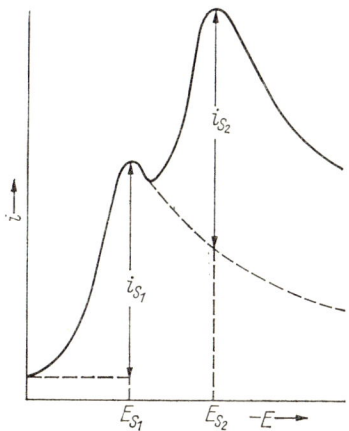

Fig. 13. Oscillogram of a mixed depolarizer. The reference current corresponding to the second current peak must be extrapolated.

A particular advantage of oscillographic polarography lies in the *rapidity* with which the polarogram is traced. This advantage is naturally important only when many determinations have to be made in sequence, because the time-controlling factor in polarographic analysis lies in the preparation and deaeration of the solution.

It is surprising that the main application of oscillographic polarography should lie in the analytical field, considering the fact, mentioned in the introduction, that problems of reaction kinetics provided the starting point for its development. It is regrettable that relatively few kinetic investigations have been published, because this is precisely where the oscillographic method fills a gap between the region of low reaction velocities to which ordinary analytical methods (e.g., titration) are still applicable, and the region of very rapid reactions, to which other electrochemical procedures (mathematical analysis of the polarographic current or current sweep; galvanostatic or potentiostatic procedures) are usually employed. Compared with the last-named methods oscillographic polarography has the advantage of greater simplicity in mathematical evaluation, because the concentrations of the participants in the reaction can be read from the polarogram. As examples mention may be made of the investigations of Favero[45] on the reaction between pyruvic acid with ammonia, and, what was probably the first work of this kind, the investigations of Snowden and Page[46] on azo-dyestuff formation by the coupling of sulfonated pyrazolone with diazotized sulfanilic acid, which gave the reaction velocity of this reaction with an error of 2 %. Figure 14 shows the

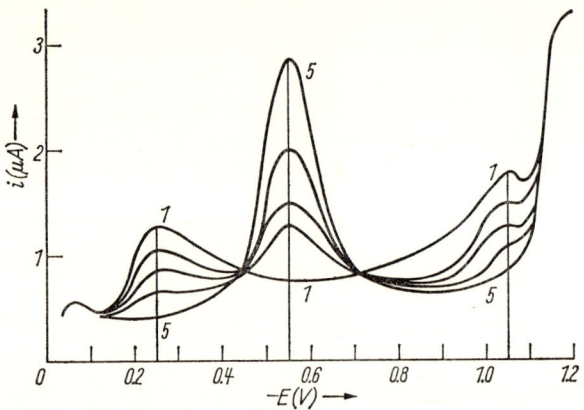

Fig. 14. Coupling of diazotized sulfanilic acid with sulfonated pyrazolone at pH 6.0 (Snowden and Page[46]). The individual oscillograms were taken 0, 15, 30, 60, and 120 sec after the start of the reaction. Left and right, the current peaks of the diazo-compound; in the middle, those of the azo-dye.

sequence of oscillograms taken 0, 15, 30, 60, and 120 sec after the start of the reaction. In the first (1) only the two peaks corresponding to the diazo-compound (left and right) are to be seen; as the reaction proceeds they become smaller, and at the end (5) disappear completely. The height of the peak corresponding to the azo-dyestuff (middle) changes in an exactly reverse manner; it is zero at the start (1) of the reaction and has reached its maximum value after 120 sec (5).

In conclusion something has to be said about the *disadvantages* of the oscillopolarographic method. First of all, its accuracy is somewhat lower, because the polarogram is relatively small, and especially be-

B. *Nonstationary Methods* 29

cause the complexity of the electronic equipment introduces possibilities of error which are very difficult to estimate. Furthermore the polarogram has usually to be photographically recorded at the expense of time and labor, and even then the picture, which is small anyhow, can be distorted though ever so slightly. Finally, the requirements in respect of equipment are very considerable, if really accurate and, above all, reproducible results are to be obtained. The last-mentioned point in particular is doubtless responsible for the halting progress of oscillographic polarography; and the position today could very well still be the same as expressed by Cruse[47]: "the natural aloofness of the chemist in matters concerning electronics has been not without influence on the development."

6. Polarography with Superimposed Alternating Voltage

The increasing significance of *alternating-current methods* in electrochemistry (impedance and admittance measurements of the most diverse kinds, measurement of dielectric constant, radio-frequency titrations, etc.) appears also in the modern development of polarography. Strictly speaking this should include all procedures which involve a voltage that varies periodically, such as the multi-sweep method described in the previous chapter, and oscillographic polarography according to the method of Heyrovský, which used the 50-cycle mains voltage, and which will be discussed later. In order to be systematic the term "alternating-current polarography" will be applied here only to those methods in which an *alternating voltage of small amplitude* is superimposed on the polarizing voltage.

a. *Conventional Alternating-Current Polarography*

The first experiments of this kind are already 20 years old and were made by Müller et al.,[48] who superimposed a sine-wave alternating voltage, and observed the alternating current on the screen of a cathode-ray tube. The latter is an exact sine-wave only if the zero line of the alternating voltage coincides with the half-wave potential (Fig. 15) and exhibits unsymmetrical distortions before and after this potential. The method ought therefore to be suitable for the determination of half-wave potentials. It is, however, necessary for the polarographic curve itself to be symmetrical.

Boeke and van Suchtelen[49,50] modified the method considerably in that they did not produce the alternating current itself but the phase

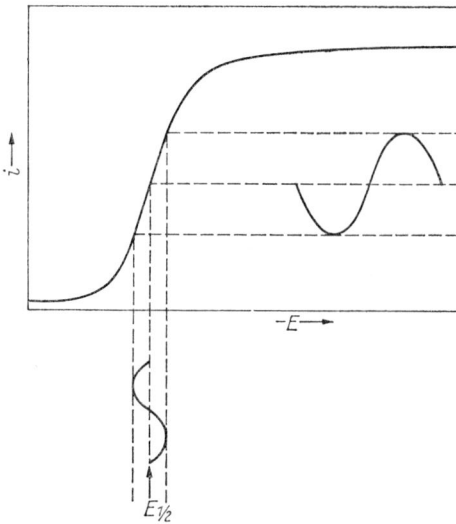

Fig. 15. Determination of the half-wave potential by superimposing a sine-wave A.C. voltage (Müller et al.[48]).

displacement between current and voltage using Lissajous figures. The phase displacement shows a minimum at the half-wave potential, and is more clearly distinguishable than the distortion of the sine wave used by Müller and his co-workers. Neither method has acquired much importance, mainly because the analyst is more interested in the height of the peak than in the half-wave potential, which usually requires correction.

The first applications of alternating-current polarography to quantitative analytical measurements were published by MacAleavy[51] and, somewhat later, by Breyer and Gutmann.[52,53] The principle of the method is shown in Fig. 16. The polarographic current circuit is no different from that used in *direct-current* polarography if one ignores the source W of alternating voltage. The linearly increasing D.C. voltage G is applied to the cell Z. The resulting direct current flows through the variable resistor R and produces a voltage drop, which in the case of direct-current polarography after suitable amplification (V) is registered by the instrument M. In *alternating-current* polarography a source W of alternating voltage is inserted in series with the applied D.C. voltage. This is usually a sine-wave having the following properties: (1) low frequency, usually between 1 and 250 cycles, (2) constant frequency, (3) small amplitude, between 1 and 50 mV, and (4) constant

B. Nonstationary Methods

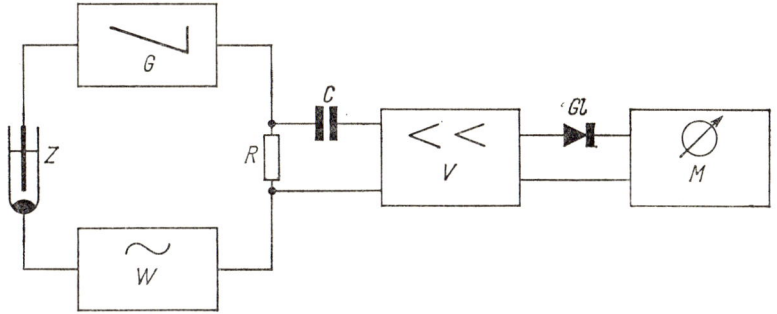

FIG. 16. Schematic diagram for alternating-current polarography.

amplitude, which does not increase with time, as does the D.C. polarizing voltage. A D.C. voltage modulated by an alternating voltage is thus applied to the cell. The current which flows is naturally greatly influenced by the superimposed alternating voltage, for it is affected not only by the pure ohmic resistance but also by the impedance of the whole circuit, and by the additional fact that the electrochemical reactions at the electrode are of a different nature from those that occur when a D.C. voltage alone is applied. The total current may be regarded as divided into a direct-current portion and an alternating-current portion. The latter only is measured simply by inserting between the resistor R and the amplifier V a blocking capacitor C, which if properly chosen will completely stop the direct current, but offer little resistance to the alternating current. The voltage to be measured is then amplified by the amplifier V, rectified by the rectifier Gl, and measured by the instrument M. The resulting polarogram shows the dependence of the *alternating current* $i_\sim$ on the applied D.C. voltage $E_=$. As shown schematically in Fig. 17, instead of the usual polarographic wave, we have a characteristic peak whose peak potential to a first approximation coincides with the half-wave potential of the direct-current polarogram. This curve form recalls that which is obtained in derivative polarography. Nevertheless, we are not dealing here with a simple derivative of the kind where the alternating voltage (as in Fig. 15) is "mirrored" in the direct-current polarogram as it would be to the reference line of an amplifying valve. That this is not the case is already evident from the fact that all the other irregularities of the polarogram (serrations resulting from the fall of drops, maxima, *etc.*) are *not* differentiated at the same time. The current maximum results rather from the electrode reaction itself, namely, through a

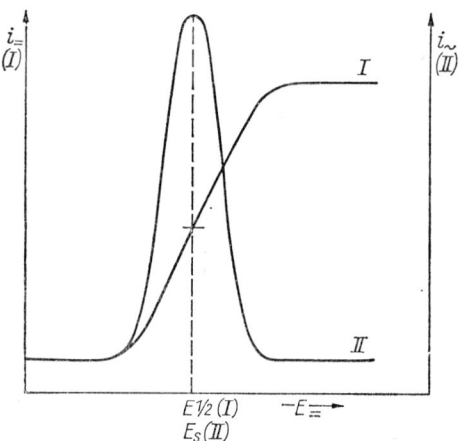

Fig. 17. Schematic representation of a polarographic D.C. wave (I) and the corresponding A.C. peak (II).

succession of electrode processes proper ("*Durchtrittsreaktionen*"), whereby the depolarizer oscillates to and fro between its oxidized and reduced state. If this interpretation were correct, the heights of the peaks would be a function of the reversibility of the electrode reaction. If the oxidation and reduction potentials are so widely separated that they are not covered by the amplitude of the alternating voltage, then this swinging backward and forward between the two states is no longer possible; there is, rather, only a minimum amount reduced during each half-wave. This provides solely a small intermittent direct current, which is considerably smaller than the peak current of a *reversible* electrode process proper. This dependence of the peak current on the degree of reversibility of the reaction does indeed occur, and will be discussed in greater detail later.

Yet another fact must be anticipated, for it likewise shows that the method described can in no way be regarded as a simple derivation from the direct-current polarogram. Not only redox reactions, but also adsorption and desorption of surface-active substances at the electrode yield characteristic peak currents in alternating-current polarography. In this case it is not a question of currents connected with the electrode processes proper, but of displacement currents, which result from the swinging backward and forward between the adsorbed and desorbed states, that causes the appearance of a peak current. This will be discussed later.

B. Nonstationary Methods

If we have placed so much emphasis on the distinction between derivative and alternating-current polarography, it is because the misleading term "derivative polarography" has unfortunately very often been applied also to alternating-current methods.

After this rather general introduction something may now be said about the theory of electrode reactions with superimposed alternating voltage. On this subject an imposing number of works has already appeared from Bauer, Breyer, Cakenberghe, Delahay, Grahame, Kambara, Koutecký, Matsuda, and Randles, to name only a few. The fact that the results obtained by different authors do not always agree, and moreover occasionally deviate considerably from the experimentally determined values, shows that we are dealing with very complicated relationships. The equations that have been deduced are not necessarily "false"; they contain rather a series of suppositions that are very difficult to justify experimentally. It will be expedient therefore to deal with these theoretical matters in somewhat greater detail here, because automatically a number of viewpoints of importance for practical application will come under discussion.

It has already been mentioned that the degree of reversibility of the electrode reaction plays an important part. First of all the discussion

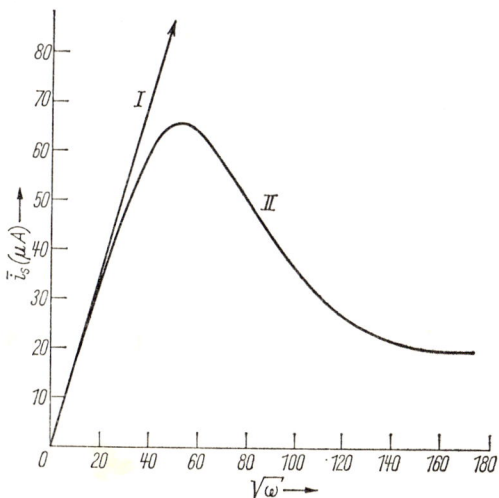

FIG. 18. Relationship between peak current and frequency. I = Theoretical according to Eq. (1); II = experimental as found by Breyer et al.[55] Solution: 10^{-3} N Tl$^+$ in 0.5 N HCl.

will be limited to *reversible* reactions. A further simplifying assumption will be made, namely, that no other changes are operative during the transport of the substances taking part in the electrode reaction except diffusion and convection resulting from the growth of the drop. In other words, we will consider first of all only *reversible diffusion currents*. With these assumptions it may be deduced that the peak current $\bar{i}_S$ (after normal rectification without regard to phase) can be represented by the following equation (Matsuda[54] and references quoted by him)

$$\bar{i}_S = knFD^{1/2}m^{2/3}t^{2/3}\omega^{1/2}c \cdot \frac{nFE_0}{2RT} \tag{1}$$

The dependence of the current on the equivalence n, the diffusion coefficient D, and the electrode surface $m^{2/3} \cdot t^{2/3}$ will not be considered further here. What are of interest to us are the new parameters appearing in alternating-current polarography, namely, the angular frequency

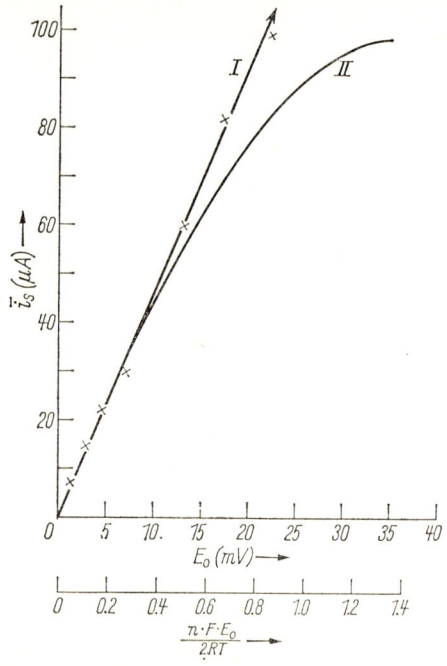

FIG. 19. Relationship between peak current and the amplitude of the superimposed A.C. voltage. I = Theoretical according to Eq. (1); II = experimental as found by Bauer and Elving.[56] Solution: $5.55 \times 10^{-4}\,M$ Cd^{2+} in $0.5\,M$ HCl. (+) plotted against $\tanh(nFE_0/2RT)$ instead of $nFE_0/2RT$.

B. Nonstationary Methods

($\omega = 2 \cdot \pi \cdot \nu$) and the amplitude of the superimposed alternating voltage (E_0), as well as, naturally, to what extent linear dependence on the concentration c is in fact fulfilled. Figures 18 to 20 show how the experimentally found figures compare with the values to be ex-

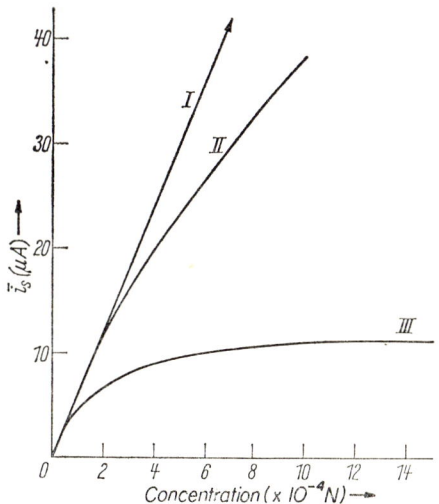

FIG. 20. Relationship between concentration and peak current. I = Theoretical according to Eq. (1); II = experimental (Cd^{2+} in 0.1 M acetate buffer, pH 4.7, total resistance 250Ω; III = same as for II, but total resistance about 2000Ω (Breyer et al.[57]).

pected from Eq. (1). It will be seen that the relationship between $\sqrt{\omega}$, E_0, or c is not linear. The reasons for this deviation are as follows:

1. Equation (1) is valid only for small amplitudes of the superimposed alternating voltage. According to Breyer et al.[55] (see also Bauer and Elving[56]) $\bar{i}_S$ does not have linear proportionality to $nFE_0/2RT$, but shows better agreement with the hyperbolic tangent of this expression (crosses in Fig. 19). In practice, however, this difference is irrelevant, as the amplitude of the alternating voltage must be chosen so small (linear portion of curve II in Fig. 19) that the electrode characteristics in the alternating-voltage range are still practically linear. For other reasons, too, it is recommended that the voltage should not be too great. For one thing the resolvability is inversely proportional to the amplitude of the superimposed voltage. If the half-wave potentials of two depolarizers lie 50 mV apart, then no

separation of the two peaks is to be expected with an alternating voltage of 50 mV; the result will be a broader collective peak. More will be said about this later.

2. Equation (1) holds exactly only when the total resistance of the polarographic current circuit is zero. Even in direct-current polarography the resistance plays a part in that it flattens the gradient of the wave and shifts the half-wave potential in the negative direction. The *height of the peak*, however, is *independent* of the resistance. On the other hand, in alternating-current polarography the amplitude of the alternating voltage applied to the electrode finds a place in the equation for the peak current. This applied voltage depends on the voltage drop resulting from the resistance of the circuit, even if the amplitude of the *source* of the alternating voltage is kept constant. For this reason the peak current *depends on the resistance*. In addition, in dealing with alternating current, the total *impedance* of the circuit plays a part as well as the ohmic resistance. This impedance may be represented as shown in Fig. 21[57] by a combination of the "faradic impedance" F, the

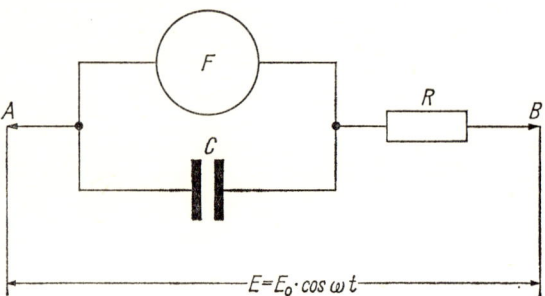

FIG. 21. Diagrammatic representation of the polarographic current circuit.

capacitance C of the electrical double layer, and the total resistance R composed of the resistance of the potentiometer, the leads, the electrodes (including the reference electrode), and the solution. Only F needs to be measured; the effect of R and C must be eliminated as far as possible experimentally or by calculation. This requirement is further complicated because phase conditions have to be taken into account in making alternating-current measurements. An ideally polarized electrode conducts as a perfect capacitor (Grahame[58]). In the absence of a depolarizer the current is determined only by the capacitance of the

B. Nonstationary Methods

electrical double layer. On the other hand, the total impedance in the presence of substances that can be reduced or adsorbed is of a more complex nature; the faradic current can not be taken as the scalar difference between total and residual current, but must be evaluated *vectorially* (Delahay and Adams[59]). As the active component of the capacitance current is always zero, the effect of the capacitance current can be eliminated experimentally by rectification which takes account of phase conditions (see Gerischer[60]); one then measures only the active component of the alternating current. As regards the wattless component one can determine the faradic current, as in direct-current polarography, by taking the scalar difference between total and residual current (Matsuda[54]).

In the case of a solution free from depolarizer, calculation shows that, with a given capacitance (C) and resistance (R), the effective voltage (E_C) is related to the applied voltage $(E = E_0 \cdot \cos \omega t)$ at points A and B (Fig. 21) as follows (Schmidt and von Stackelberg[61]):

$$E_C = \frac{E_0}{\sqrt{\omega^2 R^2 C^2 + 1}} \cdot \sin(\omega t + \varphi) \qquad (2)$$

When $R = 0$, $\varphi = \pi/2$ and $\sin(\omega t + \varphi) = \cos \omega t$. In this case the denominator $= 1$, and E_C becomes identical with E. This condition, however, is only very incompletely realized in practice. If in deducing Eq. (1), instead of E_0, we use the value found in Eq. (2), we then have for the peak current:

$$\bar{i}_S = K \cdot \frac{E_0 \cdot \sqrt{\omega}}{\sqrt{R^2 C^2 \omega^2 + 1}} \qquad (3)$$

(parameters which are not of interest to us here have been collected together in K). Figure 22 shows curves calculated for various values of R using Eq. (3). A comparison with curve II in Fig. 18 shows that the agreement with the experimentally determined curve form is most satisfactory. Equation (3) shows, moreover, another fact, first established by Kalyanasundaram,[62] namely, that the maximum of the curve giving the relationship between frequency and increasing impedance is shifted toward lower frequencies and lower values of current (curves I and II in Fig. 22).

In the considerations we have made so far, nothing has been said about the action of the depolarizer. The additional faradic impedance which it introduces does not remain constant while the polarogram is being taken, but reaches a minimum at the peak potential, when the maximum current $\bar{i}_S$ flows. The voltage gradient then has its max-

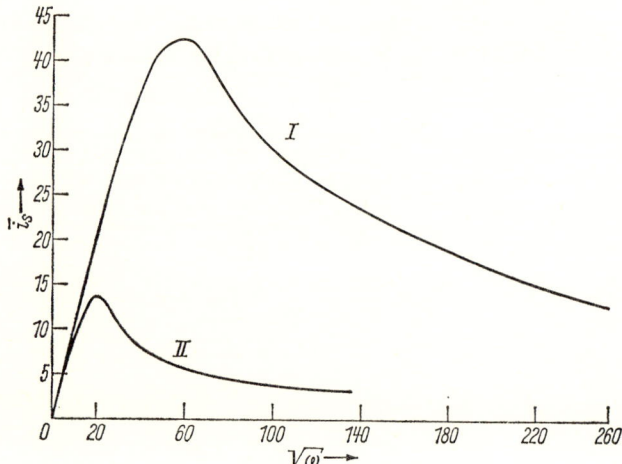

Fig. 22. Relationship between frequency and peak current calculated from Eq. (3). I, with $R = 100\Omega$, $C = 3\,\mu\text{F}$; II, with $R = 1000\Omega$, $C = 3\,\mu\text{F}$ (arbitrary ordinates).

imum value, and the effective alternating-voltage amplitude has its minimum value.

The presence of the voltage gradient has yet another important consequence. The current should increase linearly as the concentration of the depolarizer increases. For a given impedance of the circuit the voltage gradient increases as the current increases, and hence the alternating-voltage amplitude will decrease to a comparable extent. This means that for the same conductivity (the increase in conductivity resulting from a higher concentration of the depolarizer may be neglected in comparison with the total conductivity) and the same alternating voltage applied *to the cell*, the effective voltage amplitude applied *to the electrode* decreases as the concentration of the depolarizer increases. A linear relationship between concentration and peak current (Fig. 20) is less to be expected, the greater the impedance of the circuit.

The expression that can be deduced for the limiting case of a *completely irreversible electrode reaction* (reverse reaction negligible) is relatively simple. According to Matsuda[54] (again for ordinary rectification)

$$\bar{\imath}_S = k'nFD^{1/2}m^{2/3}t^{1/6} \cdot \frac{anFE_0}{RT} \cdot [1 - 1.20 \cdot (\omega t)^{-0.222}] \tag{4}$$

B. Nonstationary Methods

where α is the transfer coefficient. A condition for the irreversible limiting case is that the rate constant of the reaction which determines the potential shall be

$$k_G < 3 \cdot 10^{-5} \cdot t^{-1/2} \quad [\text{cm sec}^{-1}] \tag{5}$$

at the corresponding normal potential (t = drop-time). For the *reversible* limiting case

$$k_G > 0.15 \cdot \omega^{1/2} \quad [\text{cm sec}^{-1}] \tag{6}$$

In the reversible case the peak potential E_S coincides with the half-wave potential $E_{1/2}$ of ordinary direct-current polarography, while in the irreversible case the relation between the two is

$$E_S - E_{1/2} = -\frac{RT}{\alpha n F} \cdot [0.256 \ln (\omega t) + 0.008] \tag{7}$$

Equation (4) shows that the peak current in the irreversible case no longer increases as the square root of the frequency. We have, therefore, two factors which control the deviation of the peak current (Fig. 18, curve I) from its theoretical dependence on frequency: first, the effect of the impedance of the current circuit [Eq. (3), Fig. 22], and second, the effect of the degree of reversibility of the reaction. The two effects superimpose, and yield the frequency curve (Fig. 18, curve II) which is found by experiment.

It is not possible to obtain a simpler and more explicit expression for the peak currents of those reactions which lie between the limiting conditions given by the pairs of equations (1) and (4), and (6) and (5). Such an equation would be of little use for analytical applications of alternating-current polarography, because (1) one is restricted on account of the sensitivity, which is otherwise too low, to reactions having the greatest reversibility; and (2) the quantities k_G and α appearing in such an equation are known only in the rarest instances. On the other hand, the mathematical treatment of this problem is of significance from the point of view of reaction kinetics, as the values of k_G and α can be calculated from known concentrations of depolarizer. Such calculations have been carried out in various ways. Breyer and Gutmann[53] have deduced a corresponding equation, which is only very approximately valid, and has been recently more generally expressed by Bauer.[63] Randles[64] was able to show that the electrode reaction could be represented electrically by a resistance R_R (polarization resistance) and a capacitance C_R (pseudo-capacitance) (Grahame[65]) connected in series. By vectorial evaluation of these quantities, naturally taking into account the ohmic resistance of the circuit and the capacitance of the double layer, it is possible to calculate the velocity constant of the reaction from the values of R_R and C_R related thereto.

Van Cakenberghe[66] has deduced that harmonics of the fundamental frequency occur under conditions of alternating-current polarography which disappear in the reversible case at the half-wave potential. In the irreversible case, this potential at which they disappear (E_Z) does not coincide with $E_{1/2}$, and from the difference of these two potentials it is possible to calculate the transfer coefficient α. Van Cakenberghe, for the purpose of confirming E_Z, observed the alternating current on a cathode-ray screen as in the old method of Müller et al.[48] Bauer and Elving[67] used a frequency filter to eliminate the fundamental frequency, and so were able to read the minimum of the harmonics directly from the polarogram.

We will refrain from giving a detailed account of all these matters here, especially as the calculations are very complicated and of limited application only, and because they contain many restrictive assumptions. From the practical point of view one fact must be remembered, namely, that the height of the peak of the alternating-current polarogram is decisively influenced by the *degree of reversibility* of the reaction. One must endeavor, therefore, to have an electrode process with the greatest degree of reversibility. This can usually be accomplished to a great extent simply by a suitable choice of the composition of the solution (e.g., complex formation). On the other hand, a lowering of the height of the peak can even be of use—when it is possible, as in the analytical determination of trace substances—to reduce the interfering substance present in excess as far as possible irreversibly, and to reduce the trace component required as far as possible reversibly. The need for systematic trials in order to find the best possible composition of the solution arises here much more than in direct-current polarography. It is a fortunate circumstance that the solutions usually do not need to be deaerated, provided that their concentration is not too low, because the oxygen is reduced irreversibly. Care must be taken though, as the oxygen may participate in the reaction. For example, zinc in oxygen-free 0.1 N KCl solution gives a well-defined peak, but none at all in a solution containing oxygen.[57] In this case the OH⁻ ions formed by the reduction of the oxygen produce zincate ions, which are reduced irreversibly. Such instances are by no means uncommon.

B. Nonstationary Methods

Our considerations have so far been confined to cases involving the *electrode processes proper*; we will now deal with those in which electric charges do not *pass through* the electrode/solution phase boundary, but are only *displaced*. Curve I of Fig. 23 represents the reference current

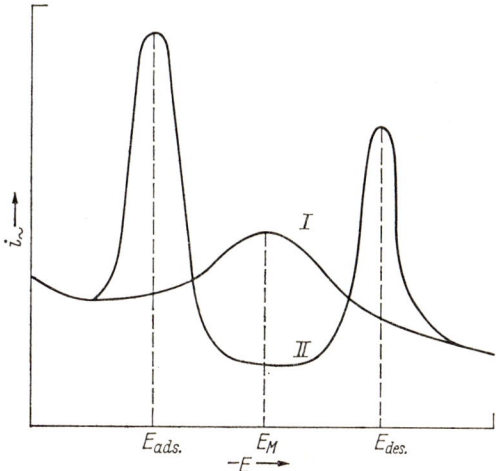

FIG. 23. Schematic representation of the A.C. polarogram of a pure supporting electrolyte (I), and the same supporting electrolyte after addition of a surface-active agent (II). E_M = Electrode zero point; E_{ads} = adsorption potential; E_{des} = desorption potential.

of an alternating-current polarogram without depolarizer. (In the presence of sufficient conducting salt this current has a maximum at the electrocapillary zero potential E_M.) If we add a surface-active substance to the solution (curve II), then the reference current is depressed where the substance is adsorbed, because the capacitance of the electrical double layer is diminished as a result of the adsorption. Furthermore, on either side of the electrocapillary zero point we find a characteristic current peak, the first at the potential where the substance is adsorbed, and the second where it is desorbed. The substance is adsorbed at potentials in the region between the two current peaks; it is desorbed outside this region. The peak currents themselves are produced because the substance oscillates to and fro between the adsorbed and desorbed states in rhythm with the superimposed alternating voltage. The curves thus provide a measure of the differential capacitance as a function of the applied voltage. Should either or both of the adsorption and desorption potentials lie outside the potential range of the polarogram, then naturally there will be one current peak only, or hardly any at all. In the latter case the reference current is lower than that of the pure solution.

Many measurements of differential capacitance for the purpose of studying adsorption changes have already been made, either using an impedance bridge, that is to say discontinuously, or continuously by means of oscillopolarographic methods. Alternating-current polarography was first used for such measurements by Doss and Gupta,[68] as well as by Breyer and Hacobian.[69] As a result of such adsorption changes the electrode remains further polarized; thus it is senseless to talk about depolarizers in the case of surface-active substances that are neither reduced nor oxidized, but do nevertheless produce a displacement current. Breyer and his collaborators therefore designated such current peaks as "tensammetric maxima."

Tensammetry may be used in analysis *to determine the concentration of surface-active substances*. As the height, the position, and the frequency dependence of the maxima are to a considerable extent regulated by the composition of the solution,[70] it is necessary in each

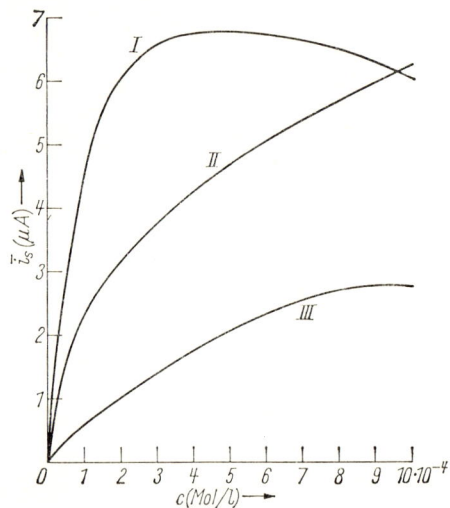

FIG. 24. Calibration curves of tensammetric maxima in a Na_2WO_4 solution (Breyer and Hacobian[71]). I: adsorption maxima in 0.1 M phosphate-borate-acetate buffer, pH 1.8; II, desorption maxima in the same solution; III, maxima in the same solution, pH 4.

B. Nonstationary Methods

case to take calibration curves. Here, too, it is frequently desirable to make trials in order to discover the most suitable composition for the solution. As an example Fig. 24[71] shows three calibration curves for Na_2WO_4: curve I for the adsorption maxima in 0.1 M phosphate-borate-acetate buffer (pH = 1.8); curve II for the desorption maxima in the same solution; curve III for the maxima in the same buffer at pH = 4 (in this case only one tensammetric maximum appears). In this example it will be possible to evaluate the first maximum down to concentrations of about $3 \times 10^{-4}\ M$, and the second at higher concentrations, for the height of the first is practically independent of the concentration. The solution with pH = 4 is less suitable for the determination of the concentration.

A particular case arises when the adsorbable substance is not previously present in the solution but appears only as a result of oxidation or reduction of a depolarizer. In this case the current maximum results from a combination of a redox and an adsorption-desorption reaction. Such maxima are characterized by being very pointed and unsymmetrical (Breyer and Hacobian[72]). The polarograms of halogen ions may be mentioned as an example. The two reactions that are superimposed here are

$$2\ Hg + 2\ Hal^- \rightleftharpoons 2\ Hg_2Hal_2 + 2e^-$$

and

$$(Hg_2Hal_2)_{ads} \rightleftharpoons (Hg_2Hal_2)_{des}$$

These peak currents are hardly suited to the analytical determination of such ions (in this case Cl^-, Br^-, I^-) when they are present *together*, because there is mutual influence on their height; *individual* determinations, on the other hand, can be made quite accurately.

Many organic compounds too give combined maxima resulting from reduction and displacement currents. These current peaks are relatively high, even when the total reaction is irreversible. Breyer and his collaborators[73] explain this fact by reversible *partial processes* comprising the reduction.

Apart from analytical applications tensammetry is suitable for the multiplicity of investigations related to *interfacial phenomena*. These matters have so far received little theoretical consideration, but the results indicate that the method certainly contains many as yet unexploited possibilities. One fresh example only need be mentioned, namely, the investigation into the adsorption behavior of organic compounds at a mercury surface by Schwabe and Jehring,[74] who ascertained among other things that organic solvents are able to displace organic depolarizers from the adsorbed layer.

The results so far discussed have more than theoretical interest; they produce a number of important requirements from the practical point of view, which may be summarized as follows:

The *amplitude* of the superimposed alternating voltage should be as far as possible *lower* than 30 mV (Fig. 19). Greater amplitudes do indeed provide higher peak currents, but in this case the electrode characteristic is no longer to be considered as linear. Furthermore, the resolvability is less.

The *frequency* of the superimposed voltage should likewise be as *low* as possible; in fact, the higher the impedance of the circuit, the lower it should be. Higher frequencies do indeed provide higher peak currents, but only up to some maximum value, which is determined by the impedance of the circuit [Eq. (3), Fig. 22] and the reversibility of the reaction [Eq. (4)]. As, moreover, the peak current increases to a first approximation proportionally to the square root of the frequency, and the capacitance current linearly with the frequency, it follows that at higher frequencies the ratio of the faradic current to the capacitance current becomes more unfavorable, and the effective sensitivity correspondingly lower. So we have here, *mutatis mutandis*, the same conditions as described in the previous chapter on oscillographic polarography (Fig. 12).

The *impedance* of the current circuit must be kept as *low* as possible. It has an effect on (1) the *position* of the current peaks (exactly as in direct-current polarography the waves are displaced when the resistance increases); (2) the effective amplitude of the alternating voltage applied to the electrode [Eq. (2)], and hence the height of the current peaks (Fig. 22); (3) the magnitude of the capacitance current; (4) the relationship between frequency and current maxima [Eq. (3)]; and (5) the relationship between concentration and peak currents [Fig. 20].

As the capacitance of the electrical double layer, which lies between about 20 and 40 μF/cm^2, is affected only to a very small extent, the requirement to reduce the impedance is practically synonymous with the necessity to reduce the ohmic resistance of the circuit. For this purpose the resistance of the leads and especially the input resistance of the amplifier must be kept as low as possible. Moreover, one should avoid, whenever possible, using a separated reference electrode with its

B. Nonstationary Methods

high diaphragm resistance. Finally, the resistance of the solution is to be reduced by a high concentration of the conducting salt.*

For exact measurements the faradic current must be treated as the *vector* difference, and not as the *scalar* difference, between the total current and the capacitance current. In practice certainly one invariably measures the peak current directly versus the reference current. In contrast with direct-current polarography the current measured is not the mean value over the whole life of the drop, but its value at the moment of falling. This need not always coincide with the maximum value of the current during the life of the drop, because the current-time curve of the drop can have a decreasing characteristic according to the value of the applied voltage and the composition of the solution.

The *composition of the solution* must be so chosen that the substance under consideration is reduced as *reversibly* as possible. On the other hand, by making the course of the reaction *irreversible*, it is possible to suppress unwanted peaks.

Surface-active substances are to be *excluded* as far as possible, unless they are to be determined by the aid of tensammetric maxima, because they not only affect the peak currents of redox reactions, but also are able themselves to give interfering current peaks.

How novel and yet important these matters are for practical workers in direct-current polarography may be shown by the following: In direct-current polarography the height of the wave is independent of the resistance of the circuit, the degree of reversibility of the electrode reaction, and the speed of the voltage sweep (if this is small compared with the drop rate). With certain exceptions surface-active substances in low concentrations have no effect on the height of the wave, and are even welcome in order to suppress streaming maxima. The faradic current is practically independent of the magnitude of the capacitance of the electrical double layer, and the capacitance current, of the magnitude of the ohmic resistance; by scalar addition the two give the total current. Finally, when one takes into consideration the fact that all the sources of error present in direct-current polarography (such as depletion effect, back pressure of the mercury, screening effect, and so on)

---

* The fact that a considerable *increase* in current can be achieved under certain circumstances by decreasing the concentration of the base electrolyte (von Sturm and Ressel[75]) originates in a quite different way, and will be discussed later (see p. 46).

must also be taken account of in alternating-current polarography, and that corresponding to the greater complexity of the necessary equipment the possibilities of errors are increased, one might conclude that alternating-current polarography is too complicated, and therefore an unsuitable method. This is by no means the case. The evaluation of peak currents for analytical purposes is always made with respect to calibration curves, as is usual in direct-current polarography. As the majority of the parameters (amplitude, frequency, etc.) are fixed within the critical limits by the apparatus employed, it is really only necessary to ensure that all the related polarograms are taken under strictly comparable conditions. For example, it is not permissible to compare two peak currents that have been observed in different solutions, or under different conditions of frequency, alternating-voltage amplitude, or resistance (as by using a separated mercury pool in one case and not in the other). If the foregoing basic rules are observed, then alternating-current polarography is neither more complicated nor less reliable than conventional polarography. We will now compare the two methods in respect of sensitivity, resolvability, and separability.

The *sensitivity* is about the same as in direct-current polarography, though only for reversible reactions, as has already been fully discussed. According to von Sturm and Ressel[75] the sensitivity in the case of irreversible reactions can be considerably increased under certain conditions, if the concentration of the base electrolyte is decreased, although the resistance of the solution is naturally increased as a result. Figure

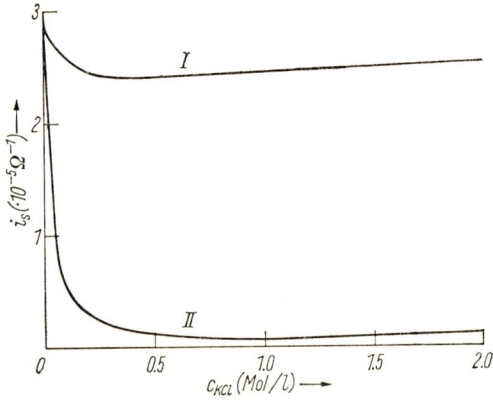

FIG. 25. Relationship between height of peak and concentration of the base electrolyte (von Sturm and Ressel[75]). I: $10^{-5}\,M$ CdCl$_2$, $10^{-2}\,M$ HCl; II: $10^{-5}\,M$ ZnCl$_2$, $10^{-2}\,M$ HCl.

B. Nonstationary Methods

25 shows the heights of the peaks given by zinc and cadmium ions in relation to the concentration of the base electrolyte.* The great increase in the height of the wave given by zinc at low concentrations is evident. It may be concluded that we are not dealing with a conductivity effect, because this increase is inconsiderable for the reversible cadmium curve, but is enormous for the irreversible zinc curve. Von Sturm and Ressel have also investigated how this effect depends on the nature of the electrolyte (specific or nonspecific adsorption of ions). They came to the conclusion that there was some effect on the structure of the electrical double layer. With diminished ionic strength in the solution a diffuse double layer is formed, which extends throughout the Helmholtz layer. The potential drop in this layer necessitates that the potential "at a distance equal to the radius of one molecule from the surface of the electrode" (ψ_1-potential of Frumkin) is no longer equal to the potential within the solution. Indeed a negative value for the ψ_1-potential on the negative branch of the electrocapillary curve results in an increase in the concentration of cations (e.g., Zn^{2+}) at the surface of the electrode. This accelerates the electrode reaction—in the case in question, probably by accelerating a reaction which precedes the electrode process proper, such as repulsion of the hydrate shell. Preliminary stages of this kind have already been suggested by Barker[76] to explain deviations from the values of peak current to be expected on theoretical grounds in the case of highly irreversible reactions (e.g., Ni and Co).

The effect described is of great significance for analytical purposes involving irreversible electrode reactions, which otherwise show up a weakness in alternating-current polarography, because the sensitivity can be increased to a very considerable degree. The resolvability and separability may be increased, for example by suppressing an interfering irreversible peak as a result of increasing the concentration of the conducting salt. Figure 26 shows an example. The cadmium peak can be observed with certainty in the presence of a 200-fold excess of chromium only when the concentration of the conducting salt is made so high that the chromium peak becomes negligibly small.

Alternating-current methods that in general have a greater degree of sensitivity will be treated in the next sections.

* The ordinates in Figures 25 and 26 do not refer to the alternating current, but to the reciprocal of the resistance, because in alternating-current polarography it is not customary to measure the absolute current, but instead the value of the resistance which, when substituted for the cell, gives the same deflection on the recorder.

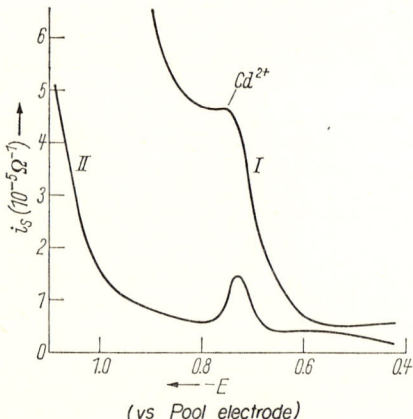

Fig. 26. Determination of Cd^{2+} in the presence of a 200-fold excess of Cr^{3+} by increasing the concentration of the base electrolyte (von Sturm and Ressel[75]). I: $5 \cdot 10^{-6}$ M $CdCl_2$, 10^{-2} M $CrCl_3$, 0.5 M Na_2SO_4. II: as for I, but with 1.0 M Na_2SO_4.

The *resolvability* is considerably better in alternating-current polarography than in direct-current polarography. This is an advantage basically common to all alternating-current methods. It has already been established, in discussing derivative polarography, that current *peaks* are generally easier to separate than current *waves*; and the defects (exaggerated serrations, deformation of curves, etc.) do not occur in alternating-current polarography. The reason for the improved resolvability is that waves in sequence do not need to be measured with reference to each other, because each by itself is related to the common reference current. The difficulty with direct-current polarography is precisely this, that waves following closely together merge without leaving any clearly defined limiting current or reference current between them. It is true that two alternating-current peaks may merge if they are very close together without the current falling down to the reference current in between, but this intermediate value is not critical provided that the two peaks are clearly defined and can be measured with respect to the reference current (with the restriction that a scalar measurement is only an approximation, though a very good one). Figure 27 gives as an example the alternating-current polarogram of various depolarizers compared with their direct-current polarogram. In the alternating-current polarogram, Pb in the presence

B. Nonstationary Methods

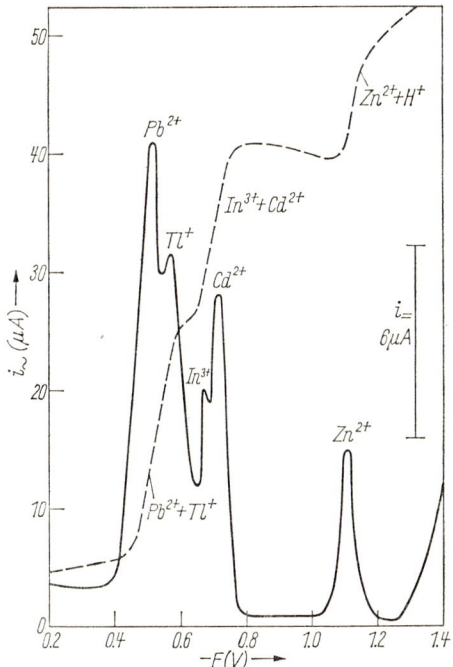

FIG. 27. ——— A.C. polarogram, — — — D.C. polarogram. Pb^{2+}, Tl^+, In^{3+}, Cd^{2+}, and Zn^{2+} each 10^{-3} N in 0.1 N HCl. $E_0 = 15$ mV (Breyer et al.[57]).

of Tl and In in the presence of Cd may be easily identified, while in the direct-current polarogram they merge in the same wave.

It has already been mentioned that the resolvability is better, the smaller the amplitude of the alternating voltage. This is shown by the example in Fig. 28.* As the amplitude becomes smaller, the current falls; so a compromise has to be made between high resolvability and high sensitivity.

Finally a figure may be given as a comparison: In the alternating-current polarogram two depolarizers may still be determined in the presence of each other when their peak potentials differ by 40 mV or more.

It is easy to see that the *separability* in alternating-current polarog-

* The polarograms of Figs. 27 and 28 were obtained by Breyer and his collaborators[55] polarometrically, that is to say, point by point. For the sake of clarity the individual points are not shown. Note the different scales for ordinates in Fig. 28.

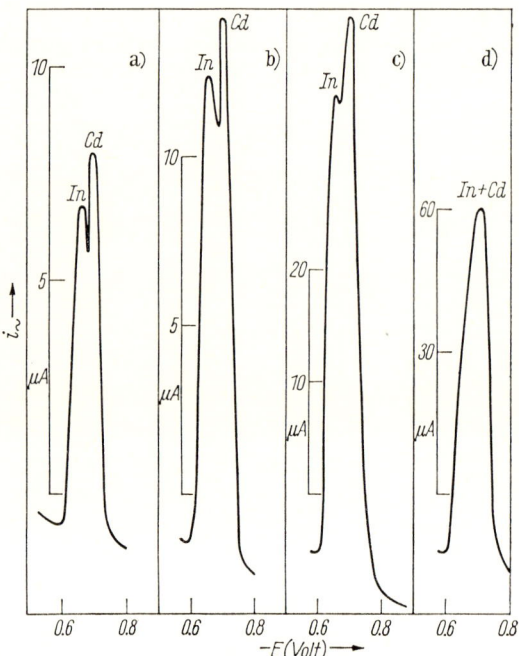

FIG. 28. Resolvability in relation to the amplitude of the superimposed A.C. voltage (Breyer et al.[57]). Solution: 10^{-3} N Cd^{2+}, 1.33×10^{-3} N In^{3+} in 0.1 N HCl. $E_0 = 5$ mV (a); 10 mV (b); 30 mV (c); 45 mV (d).

raphy is considerably better too. For if the currents of the individual redox reactions are not additive, as in direct-current polarography, but each one is formed from the reference current, then the size of the serrations caused by the falling of the drops must be independent of those reactions which take place *prior* to the arrival of the peak to be measured, that is to say, at a more positive potential. In alternating-current polarography two substances may still be determined in each other's presence if the concentration ratio is 1:100, and in favorable cases if it is 1:1000 or more, no matter which of the two is reduced at the more positive potential. Naturally the separability is much smaller if the two peaks lie very close together, because the current maxima also become broader as the height increases.

Other advantages that may be mentioned are these: the possibility of being able to determine surface-active substances, and to influence the height of the peak by a suitable choice of composition of the solution

B. Nonstationary Methods

(degree of reversibility). This can also be a disadvantage, as the method has only limited applicability for substances that are irreversibly reduced. There is a wide field of application in the investigation of electrode reactions. It is clear that the possibilities of obtaining information on the changes at the electrode-solution phase boundary and in the electrical double layer are multiplied. A combination of direct-current and alternating-current investigations has proved particularly fruitful, for it must be emphasized here that alternating-current polarography is in no way a *substitute* for, but a welcome *supplement* to direct-current polarography.

An excellent example of this is the investigation of cystine by Biegler and Breyer[77] in which a clear picture of the course of the reaction was secured by an intelligent interpretation of the various reversible, irreversible, kinetic, and tensammetric waves obtained in the presence and absence of surface-active substances in the direct-current and alternating-current polarograms.

Apart from the limited applicability of the method, as already mentioned, to irreversible processes, two final disadvantages of alternating-current polarography are the greater amount of equipment that is needed, and the greater number of possibilities of error.

b. Square-Wave Polarography

Square-wave polarography was developed by Barker and Jenkins[78] with the object of establishing a method having the advantages of alternating-current polarography, but with an increased sensitivity. It was stated in the introduction that the capacitance current must somehow be eliminated from the current that is being measured if the sensitivity is to be increased; also in alternating-current polarography there is a capacitance current produced by the periodic changes in the charge of the electrical double layer in addition to the main capacitance current resulting from the growth of the drop. Hence, square-wave polarography must take account of these two components of the capacitance current. This constitutes the main function of the square-wave polarograph.

Initially it worked like an ordinary alternating-current polarograph, only with the difference that it employed *rectangular* instead of sine-wave alternating voltage, for reasons that will be given later. Fundamentally it was no different from the alternating-voltage method, as each rectangular pulse of voltage may be split up into the sum of sine-wave voltages by means of Fourier analysis. The frequency of a commercial apparatus* is 225 cycles. This value was chosen (1) in order to obviate interference from the mains frequency and its harmonics; (2) to work in a region where the linear increase in the frequency curve (Fig. 22**) is still operative with values of resistance ordinarily encountered in the polarographic current circuit; and (3) to have a favorable ratio of time scale to the time constant of whatever is being measured. This point will now be considered in detail.

The following artifices have been employed to eliminate both of the above-mentioned components of the capacitance current.

1. The capacitance current resulting from the growth of the drop is eliminated by measuring the current only at the end of the drop life, when the increase in the surface of the electrode in unit time is quite small. The same principle is used in strobe polarography, where the matter has already been treated in detail, so it is not necessary to make a repetition here. The small capacitance current resulting from the increase in surface, which is still flowing even shortly before the fall of the drop, is compensated by a linear opposing voltage using the old method of Ilkovič and Semerano.[1] However, in alternating-current polarography there is no linear voltage increase; the voltage is modulated periodically by the alternating voltage, which in the present case

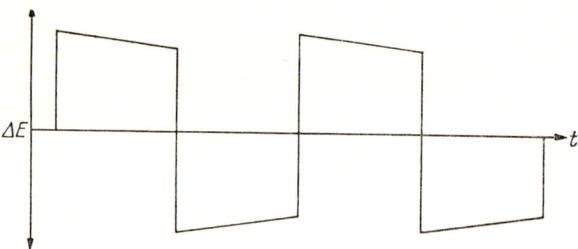

FIG. 29. Tilting the roof of a square wave to compensate the residual capacitance current (schematic).

is a square-wave voltage. Consequently the voltage employed for compensation must be modulated in like manner. This is achieved, as shown rather exaggeratedly in Fig. 29, by slightly tilting the roof of the square-wave.

* Made by Mervin Instruments, Woking, Surrey, England.
** Figure 22 is valid for sine-wave alternating voltage, but may be applied approximately to rectangular waves.

B. Nonstationary Methods

2. The elimination of the capacitance alternating current may be explained by reference to Fig. 30. Figure 30a represents the periodic course of the superimposed square-wave voltage, ignoring the tilt shown in Fig. 29. Figure 30b shows the course of the reaction current $i_{F\sim}$ on the same time scale. It rises almost vertically and then falls away until, at the reversal of the voltage, it jumps in the opposite direction. Figure 30c shows the course of the capacitance current $i_{K\sim}$. It likewise rises steeply and falls exponentially to zero, in fact much quicker than the faradic current (how quickly depends on the properties of the circuit, and will shortly be discussed more closely). The way in which the capacitance current may be eliminated is now apparent. If instead of measuring the current throughout the whole period of the square-wave impulse, one chooses a time interval (τ in Fig. 30) in which the capacitance current has already fallen practically to zero, while the faradic current still retains an appreciable value, then one does in fact succeed in obtaining a total current which is almost free from the capacitance current. Figure 30 also shows why a square-wave voltage is used.

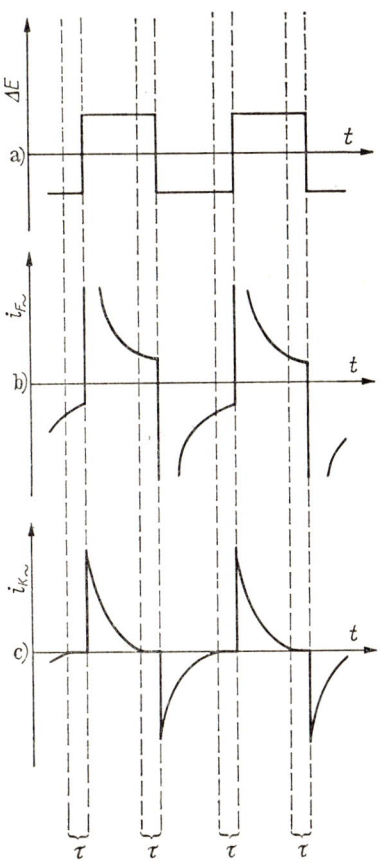

FIG. 30. Schematic representation of the periodicity of the superimposed square-wave voltage (a), the faradic current (b), and the capacitance current (c).

If the voltage changed continuously, as with a sine wave, then the course of the capacitance current and the reaction current would exhibit no sharp turning points. The method which has been described is therefore applicable only when a voltage that changes *discontinuously* is employed. Kalvoda et al.[79] have used the same principle with the commutator method of

Kalousek[80]; it finds application also in pulse polarography, which will be discussed later. A linear counter-current compensation as in Fig. 29 using sine-wave voltage would be quite complicated, whereas the production of the tilt on the roof of a square wave impulse requires only a suitably designed R-C unit.

The quantitative treatment of the conditions illustrated in Fig. 30 are as follows:

The charging current of a capacitor varies with time according to the equation

$$i = \frac{E}{R} \cdot \exp - \frac{t}{RC}$$

where E is the applied voltage, C the capacitance, and R the ohmic resistance of the circuit. When applied to square-wave polarography C would be the capacitance of the electrical double layer, and E the amplitude of the superimposed square-wave voltage. We can therefore calculate at what time, starting from the beginning of the square-wave impulse, the charging current becomes negligibly small, that is to say, has reached about 1 % of its maximum value. This time, calculated from $\ln (1/100) = -t/RC$, comes to about $5RC$. Using a frequency of 225 cycles, which gives a maximum available time of 1/450 sec, and a maximum value of 4 μF for the capacitance of the electrical double layer at the surface of the drop, it may be calculated that R, the total resistance of the polarographic circuit, must be less than 100Ω, if the contribution of the capacitance current to the total current is to be negligibly small (Schmidt and von Stackelberg[61]). It follows that it is even more imperative to have the lowest circuit resistance possible in square-wave polarography than in alternating-current polarography. This requirement has far-reaching consequences. It means that the concentration of the conducting salt should as far as possible be not less than 0.1 or, still better, 1 mol/liter. As a result, all the impurities in the conducting salt, electrolytic as well as surface-active, are present in a correspondingly higher concentration. With the extraordinarily high sensitivity of the square-wave polarograph this can give rise to troublesome interference with the reference current, even to the extent that current peaks of real interest can no longer be resolved from the background of the interference.

In this connection the following may be mentioned: At high sensitivities a capillary effect becomes noticeable; Barker[8] calls it "capillary response." It is caused by a thin liquid film penetrating into the tip of the capillary, and altering the time constant (RC) of the whole

B. Nonstationary Methods

system, which as we have seen is responsible for the efficiency of elimination of the capacitance current, and hence the height of the reference current. It is hardly possible to deal with this effect quantitatively either by theory or by measurement, because it depends on the type, dimensions, pretreatment, and age of the capillary, apart from the voltage at any moment (referred to the electrocapillary zero point) and the composition of the solution; it varies also from drop to drop. The resulting fluctuations in the reference current are more or less static, and determine the limits of sensitivity of square-wave polarography, because experience has shown that these fluctuations are about 4 times as great as the noise level produced in the reference current by the electronic equipment, which really decides the maximum sensitivity. This effect is naturally not peculiar to square-wave polarography; on the contrary it occurs in all alternating-current methods as soon as a certain sensitivity has been reached. A noticeable decrease in interference from this source may be achieved by suitable pretreatment of the capillaries, for example, by making the bottom end spherical (Barker[76]). Treatment with silicone often brings about a considerable improvement in the wave-form of the reference current, but capillaries that have been treated with silicone cannot be used in acid solution at voltages more negative than about −0.9 volt.

The requirements of square-wave polarography in respect of apparatus must be briefly outlined. (A rather more detailed description of the very complicated apparatus will be found in the memoir of Schmidt.[2]) Figure 31 is a greatly simplified schematic diagram. The linearly increasing direct-current voltage is obtained electronically from the saw-tooth generator (I); the square-wave voltage from a

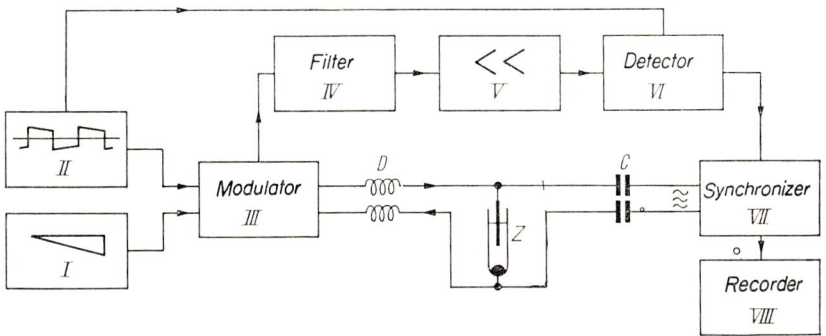

FIG. 31. Schematic diagram for square-wave polarograph.

square-wave generator (II), which also produces the tilting shown in Fig. 29. The two voltages are combined in the modulator (III), and then applied to the cell (Z). The resulting current then passes to a filter (IV), which holds back all low-frequency components, especially the direct current; it next goes to the amplifier (V), and finally to the detector (VI). The function of the detector is to leave a free path only during an interval of time τ (see Fig. 30), namely 1/8 of the total impulse time, and to block the path during the remainder of the time. The detector must also be synchronized with the square-wave generator. Moreover, the current is now rectified and smoothed. The synchronizing unit (VII) finally ensures that the measurement takes place only at the end of each drop life, namely, 1.75 sec from the beginning. For this purpose a radio-frequency voltage, free from direct-current voltage, is applied to the cell (Z) through the blocking capacitors (C) [the chokes (D) prevent the radio frequency from entering the measuring circuit proper]; and the reaction of the sudden change in impedance when the drop falls away is utilized to coordinate the drop time and the interval to be measured. The current having thus been synchronized with reference to the end of the square-wave impulse and the end of the drop life finally reaches the recorder (VIII).

The changes taking place at an electrode with superimposed square-wave voltages have been frequently treated theoretically (Kambara,[81] Koutecký,[82] Matsuda,[54] and especially Barker and his collaborators[83]). In the case of a reversible electrode reaction the amplitude of the alternating-current component is given by

$$i = \pm \frac{n^2 F^2}{RT} \cdot c \cdot A \cdot \Delta E \cdot \frac{P}{(1+P)^2} \sqrt{\frac{D}{\pi \cdot \tau}} \cdot \sum_{m=0}^{\infty} (-1)^m \cdot \frac{1}{\sqrt{m+\beta}} \quad (8)$$

where $\beta = t/\tau$ $(0 < t < \tau)$, and $P = \exp{(E - E_{1/2}) \cdot n \cdot F/RT}$. ($t$ = time [sec] from the last change in direction of the square-wave impulse, τ = half-period of the square-wave voltage [sec], E = mean potential of the mercury drop, $E_{1/2}$ = half-wave potential, c = concentration of the depolarizer [mol/cm^3], ΔE = amplitude of the superimposed square-wave voltage, A = surface area of the electrode [cm^2], and D = diffusion coefficient.) On substituting values specific for a commercial apparatus we have

$$i = 5.61 \cdot 10^7 \cdot \Delta E \cdot n^2 \sqrt{D} \cdot c \cdot A \cdot \frac{P}{(1+P)^2} \quad (9)$$

Barker has deduced corresponding equations also for the cases of the

B. Nonstationary Methods

slightly irreversible and semi-irreversible reaction. We will consider only the case of the completely irreversible reaction. For this

$$i \approx \pm \frac{n^2 F^2}{RT} \cdot A \cdot \Delta E \cdot \alpha_1 \cdot c^0 \cdot kf \sum_{m=0}^{\infty} (-1)^m \exp N^2(m+\beta) \operatorname{erfc} N\sqrt{m+\beta} \quad (10)$$

where $N = k_f \cdot \sqrt{\tau/D}$, α_1 = transfer coefficient of the forward (reduction) reaction, c^0 = concentration of the depolarizer at the surface of the electrode (this can be calculated from an equation of Koutecký[84]), and k_f = rate constant of the forward reaction. On comparing the two equations quantitatively it is seen that the completely irreversible wave has a much greater width at the mid-value, and about 1/20 of the height of the reversible wave (again using figures for the commercial apparatus). According to Barker[76] there is agreement between theory and experiment for all these equations. Nevertheless in analytical work calibration curves will always be used. All the restrictions, such as small amplitude, low frequencies, and low ohmic resistances, mentioned in connection with alternating-current polarography naturally apply here as well. Everything that was said in the previous section about the characteristics, applicability, interfering factors, and limitations of alternating-current polarography are valid also for square-wave polarography. Its distinguishing feature lies in the greatly increased *sensitivity*. According to Barker,[76] for reversible electrode reactions it goes down to 2×10^{-7} M or lower in favorable cases; and for irreversible reactions down to 10^{-6} M. With good recorders an accuracy of 0.2 % should be obtained at a concentration of 2×10^{-5} M. Ferrett et al.[85] have made investigations by different methods, and found that under comparable conditions the square-wave polarograph is 200 times more sensitive than an ordinary polarograph, 12 times more sensitive than an ordinary alternating-current polarograph, and 6 times more sensitive than a multi-sweep oscillographic polarograph. Finally, von Sturm and Kaukewitsch[86,87] have carried out comprehensive and careful work on the limits of applicability of square-wave (and direct-current) polarography. By using the three-electrode method of Vlček[88] it is possible to determine peak potentials accurately to ±2 mV. The limit of usefulness lies at 0.5×10^{-7} M for reversible, and 5×10^{-6} M for irreversible reactions. It is worth noting that 0.5-ml solution in a micro-cell suffices for a reliable measurement. Then it would be possible to determine analytically even 10^{-7} to 10^{-8} gm of depolarizer according to the degree of wave formation. In a capillary cell described by von Sturm[89] it is possible to work with 5 μl.

Resolvability and *separability* should be the same as in conventional alternating-current polarography. On account of the greater sensitivity, which permits a lower absolute concentration of the trace component, a limiting concentration ratio of 1:20,000 is reached provided that adjacent peaks are not too close.

The disadvantages of square-wave polarography are the large amount of necessary equipment, and the high concentrations of conducting salt required to lower the ohmic resistance as mentioned above. In this connection the discoveries of von Sturm and Ressel[75] relating to the increase of the peak current in the case of irreversible reductions by lowering the concentration of the conducting salt are of particular significance. This effect is indeed not specific for square-wave polarography, and has already been discussed in a previous section. It operates, however, most advantageously with the high sensitivity of the square-wave polarograph, because it not only increases the height of the peak current but also provides a much smoother reference current as the *impurities* in the conducting salt are lessened.

c. *Alternating-Current Bridge Polarography*

Alternating-current bridge polarography was developed by Takahashi and Niki[90] in Japan. Figure 32 shows the principle.* The four arms of the bridge are constituted by the resistances R_1 and R_2, the combination R_S and C_S, and the polarographic cell (R_Z, C_Z) with the source of direct-current voltage (G). The assembly is fed through C_0 and R_0 with a sine-wave alternating voltage of mains frequency with small amplitude from the source W of alternating voltage. While the bridge is out of balance there is a potential difference between the points A and B, and a current flows through the primary winding of the transformer Tr; after leaving the amplifier V it controls one phase of the balancing motor M. The other phase is fed directly from the mains. The purpose of the combination R_0 and C_0 is to displace the phase of the bridge voltage so that there is a 90° phase difference between the output voltage of the amplifier and the mains voltage. The sliding contact of the potentiometer P is coupled to the balancing motor, which operates until the bridge is balanced, which condition is satisfied when

$$\frac{R_2}{R_1} = \frac{R_S}{R_Z} \tag{11}$$

$$C_Z \cdot R_Z = C_S \cdot R_S \tag{12}$$

* A direct-current bridge polarograph working on the same principle was also constructed by Takahashi and Niki.[91]

B. Nonstationary Methods

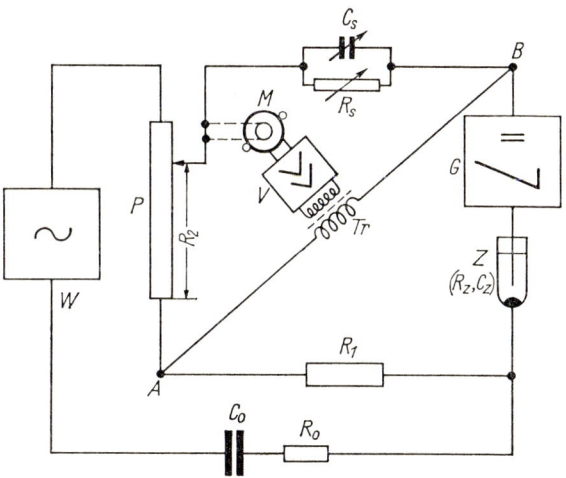

Fig. 32. Schematic diagram for A.C. bridge polarograph.

If the direct-current voltage applied to the cell rises in the usual manner, the pen recorder coupled to the sliding contact of the potentiometer P continuously registers the impedance of the cell, for R_2 according to Eq. (11) is proportional to $1/R_Z$. It is sufficient if the validity of Eq. (12) is approximately maintained by simultaneous adjustment of the resistor R_S and the capacitor C_S. No significant error is introduced into the result by this procedure.

The advantages of this novel circuit are the following: First of all, it is very simply arranged. When one remembers that compensation recorders are almost always used at present for recording polarograms, this circuit is seen to be simpler even than that of a conventional direct-current polarograph. The auxiliary voltage source is absent, and there is no direct-current amplifier; a simple alternating-voltage amplifier only is required. Miller[92] has devised a circuit which enables a conventional direct-current polarograph with compensation recorder to be used also for recording alternating-current polarograms in the following way: Alternating voltage is applied to the potentiometer supplying the compensation voltage, and the detector amplifier is used as an alternating-voltage amplifier by short-circuiting the interrupter. Even this alternating-voltage amplifier sets its problems in normal alternating-current polarography, because its input resistance is part of the total resistance of the polarographic-current circuit, which should be kept as small as possible. On the other hand, in bridge polarography hardly any current at all flows through this resistance (the primary winding of the transformer Tr) when the bridge is balanced. Consequently no voltage drop occurs there; the direct-current voltage drop over G, Tr, R_1 and Z can, however, be kept very low, and moreover is not so critical.

The main advantage of bridge polarography, as in all bridge circuits, is that a constant voltage is not necessary; the results are thus independent of the amplitude of the superimposed alternating voltage,* provided this is sufficiently small, in fact so small that the electrode characteristic can still be considered to be linear.

Sensitivity, resolvability, and separability are about the same as in ordinary alternating-current polarography. A special circuit devised by Oka[93] permits a considerable compensation of the capacitance current, with a corresponding increase in the sensitivity. We will explain this by reference to Fig. 33, considering first of all only the bridge, which is in balance, and in which the cell is represented by the substitute components R_Z and C_Z. The current flowing through R_1 is not in phase with the voltage E applied to the bridge on account of the capacitance C_Z; and in addition the current i_C flowing through C_Z is 90° out of phase with respect to the voltage E_Z applied to C_Z. We now let a current equal to

$$I_K = K \cdot E_Z = K \cdot I' \cdot Z \tag{13}$$

flow through R_1, it too being 90° out of phase with respect to E_Z; Z is the impedance of the cell, and

$$Z = \frac{1}{1/R_Z + j \cdot \omega \cdot C_Z} \tag{14}$$

For a definite value of K there is a condition of the bridge in which there is no phase difference for the voltages E, E_Z and $E_{R_1+R_2}$. The required condition may be found by a simple vectorial calculation, which we will not give, and is determined by

$$K \cdot R_1 \cdot R_Z + (R_1 + R_2)j \cdot \omega \cdot C_Z \cdot R_Z = 0 \tag{15}$$

When $R_Z \neq 0$

$$K = -j \cdot \omega \cdot C_Z \cdot \frac{R_1 + R_2}{R_1} \tag{16}$$

and the compensation current

* This is strictly true only for *linear* resistances; but the requirement is, however, satisfied by the very limited range of amplitude of alternating voltage used under practical conditions.

B. Nonstationary Methods

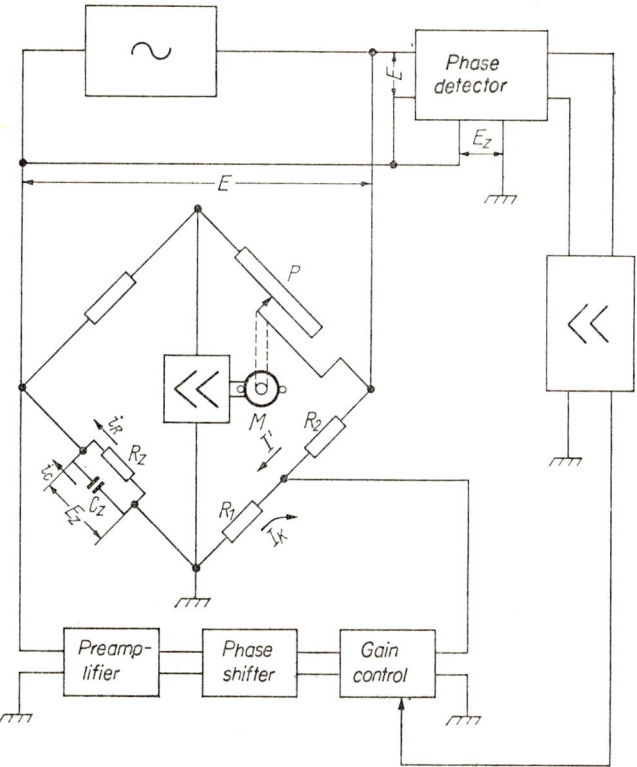

FIG. 33. Schematic diagram for A.C. bridge polarograph with capacitance-current compensation.

$$I_K = -j \cdot \omega \cdot C_Z \cdot \frac{R_1 + R_2}{R_1} \cdot \frac{R_Z \cdot E}{R_Z + R_1 + R_2} \qquad (17)$$

In practice Eq. (15) is realized as follows (see Fig. 33). The voltage E_Z is fed through a preamplifier of very high input impedance to a phase shifter, where it is displaced by 90°. The signal then goes to an automatic gain-control circuit, which supplies the voltage across the resistor R_1 where the compensation current I_K is produced. The voltages E and E_Z are compared in a phase-sensitive detector in respect of their phase relationship. This detector produces a signal which is directly proportional to the phase difference of the two voltages in question. After amplification the signal reaches the automatic gain-control circuit where it is decided what fraction of the compensation voltage must be applied to the resistance R_1 in order to satisfy Eq. (15).

The influence of the capacitance current is thus eliminated, and by means of the potentiometer P, R_Z is determined as with an ordinary conductivity bridge.

The method described is certainly very elegant, but has two disadvantages: (1) A large amount of electronic equipment is needed, even though it is not so much as for a square-wave polarograph; and (2) the capacitance current compensation is not complete, because the circuit used in replacement of the cell (Fig. 33) does not properly reproduce its true impedance (see Fig. 21). Nevertheless the increase in sensitivity is quite considerable, and is 10 to 100 times greater. For comparison two polarograms are shown side by side in Fig. 34; (a) was taken

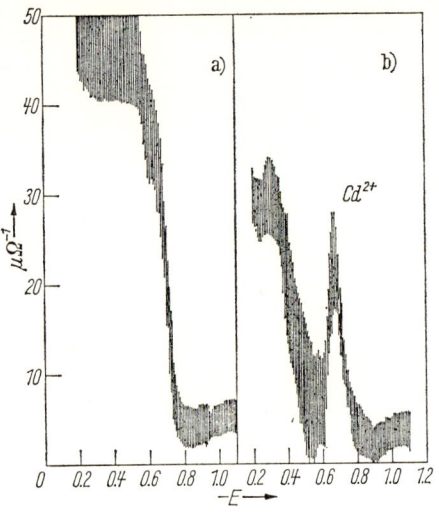

Fig. 34. Effect of capacitance-current compensation in a bridge polarograph. Solution: 10^{-5} M Cd^{2+} in 1 M KCl, without (a) and with (b) compensation. (60 cycles, 50 mV.)

without, and (b) with, counter-current compensation as described. It is obvious that without compensation the cadmium wave cannot even be distinguished qualitatively, as it is completely merged with the reference current. The example also shows that compensation is particularly necessary in dealing with peak currents lying in the region of the electrocapillary zero potential, because the reference current has a high maximum at that point.

7. Pulse Polarography

First of all, we will go over once more the principle of eliminating the capacitance current in square-wave polarography (Fig. 30). The ratio i_F/i_K is more favorable, the longer the duration of the square-wave impulse. The quantity of electricity supplied per impulse is, in respect of the capacitance current i_K, independent of the duration of the impulse (provided that the charging is complete); in respect of the faradic current i_F, it increases with the duration of the impulse. The current which is measured naturally depends on the position and duration of the time interval in which the measurement is made. In square-wave polarography the impulse time is 1/450 sec, and the time of measurement (τ) only 1/3600 sec. Better results would be obtained by increasing the duration of the impulse, that is to say, by reducing the square-wave frequency. The result would be that during the life of a drop, of which only the last portion is used in measurement, very few square-wave impulses would be received. Barker and Gardner[8,94] have therefore developed a new method, which operates so that only one square-wave impulse having a duration of 1/25 sec is applied during the life of a drop; this method is considerably different from square-wave polarography.

In pulse polarography there are basically two different ways of working. The one gives polarograms similar to those met with in direct-current polarography; the other gives polarograms similar to those in derivative polarography. Figure 35 shows the periodicity of

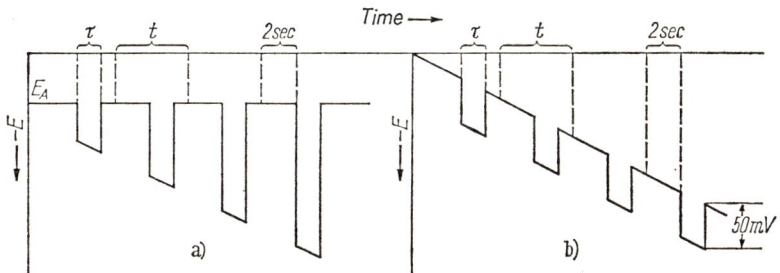

Fig. 35. Periodicity of the polarization voltage in pulse polarography. The duration τ of the impulse is 1/25 sec; the distance between two consecutive impulses is equal to the drop time t; the retardation time is 2 sec. The drawing is not to scale.

the voltage applied to the cell. Each drop receives an impulse of 1/25 sec duration, and this is applied 2 sec after the drop starts to grow (Fig. 35 is not to scale). In recording an ordinary polarogram (a) one superimposes the impulses of a constant direct-current voltage E_A (variable from about 0 to -2 V) during the recording. The amplitude of the impulse increases proportionally with time. The current resulting from the applied polarization voltage is made up as follows: There is, first of all, the reference current, which is produced by the initial voltage (E_A). All electrode reactions which are already taking place before the start of the impulses contribute to this reference current. Then there is the capacitance current, which is controlled by the charging and discharging of the electrical double layer during the course of each impulse. Finally there is a faradic current, which is provided by each impulse in the presence of the appropriate depolarizers. This current gives the variation in the diffusion current during the impulse concerned, and when related to the corresponding electrode voltage has an appearance similar to an ordinary direct-current polarogram; this change in diffusion current alone is recorded.

The capacitance current is eliminated in almost the same manner as in square-wave polarography. The compensation is more favorable, however, on account of the length of the impulse. It is measured during the second half of the impulse. The decay time for the capacitance current amounts to 1/50 sec and the period of measurement, also 1/50 sec, compared with 1/500 sec and 1/3600 sec respectively in square-wave polarography. Pulse polarography is therefore less sensitive than square-wave polarography for large time constants of the polarographic current circuit and sudden changes in these time constants (capillary effect, see page 54).

In order to eliminate the reference current a high-pass filter is used; this allows the desired current from the electrode reaction to pass, but retains the reference current of lower frequency. Dynamic filters are used, because ordinary filters are most unsuitable for very low frequencies. The measured current is thus free from all components originating in reactions taking place before the start of the voltage from the impulse. In this way we obtain polarograms of the conventional kind having current waves instead of current peaks, and have available a means of separation otherwise achieved only in alternating-current polarography.

In the second way (Fig. 35b) of taking pulse polarograms, impulses of constant amplitude (50 mV) are superimposed on a direct-current

B. Nonstationary Methods

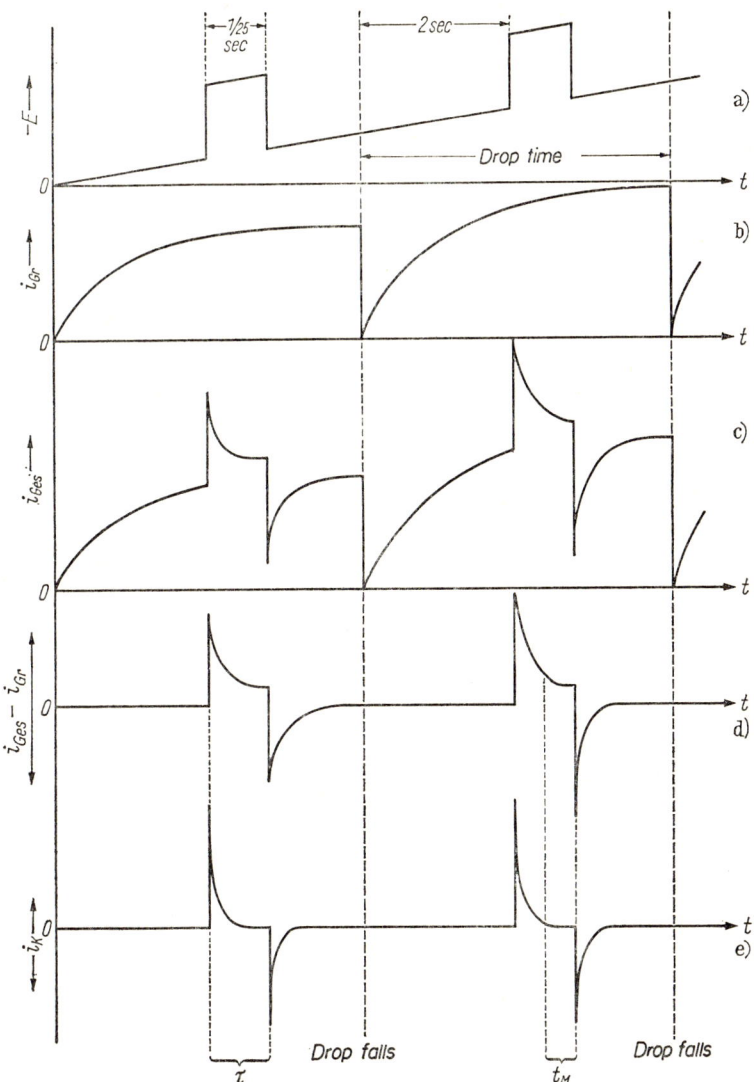

FIG. 36. Periodicity of the cell voltage and the individual current components. Details in the text.

voltage, which increases slowly in the usual manner in a negative direction. In this method also, the capacitance current and the background current, which in this case behave exactly as in an ordinary direct-current polarogram, are eliminated in the manner described above. The electrode reaction current, which is the only one required for measurement, gives the change in the diffusion current; it has a maximum, however, at the half-wave potential, because the initial voltage of each impulse is different. We obtain, therefore, a polarogram which exhibits current peaks instead of reduction waves; the result is similar to that from derivative and alternating-current polarography.

The conditions described above are shown schematically in Fig. 36: (a) shows the periodicity of the polarizing voltage (not to scale); (b) of the background current; (c) of the total current comprising the background current together with that part of the current contributed by the impulse; (d) of the current after eliminating the background current by means of a high-pass filter; and finally (e) of the capacitance current. This last is eliminated in the same way as in square-wave polarography by making the measurement during a period of time (t_M) of the second half of the total impulse time (τ) when the capacitance current has already fallen almost to zero.

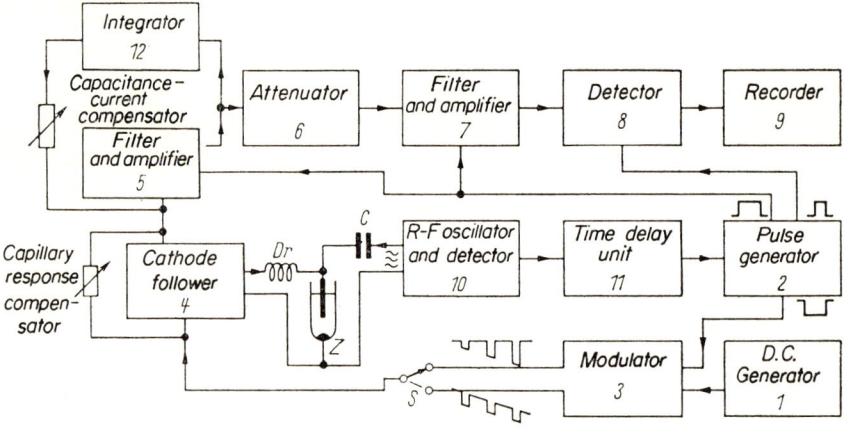

Fig. 37. Schematic diagram for pulse polarograph.

It is obvious that pulse polarography requires a very large amount of equipment. The following description can give no more than a simplified picture of a pulse polarograph* (Fig. 37). The direct-current

* Made by Southern Analytical Limited, Camberley, Surrey, England.

B. Nonstationary Methods

voltage is produced by the saw-toothed generator (1); the square-wave impulse by the impulse generator (2); the two are combined by the modulator (3). The switch S enables the circuit for producing either current waves or current peaks to be selected. The polarizing voltage (Fig. 35a or b) is applied to the cell Z through a cathode ray follower (4). The total cell current is amplified, and goes to a high-pass filter (5) which blocks the background current. The sensitivity is controlled by an attenuator (6). Then follows another dynamic filter and amplifier (7). The main function of the detector (8) is to allow passage of the current which is to be measured only during the second half of the impulse. The current after smoothing is registered by the recorder (9). To synchronize the fall of the drop a radio-frequency voltage of 10^6 cycles is applied to the cell from the oscillator (10) through the blocking capacitor C, which prevents the passage of direct current; the choke Dr prevents the high frequency from entering the circuit in which the measured current flows. The sudden changes in the impedance of the cell when the drop falls actuate a time delay circuit (11), which after the pre-set delay time of 2 sec releases the impulse for the next drop.

In order to make use of the maximum sensitivity, one must pay attention to the following points: The capacitance current resulting from the growth of the drop changes when the voltage at the electrode is altered by the impulse. This change in current is proportional to the charge that must be supplied to the electric double-layer capacitance in order to bring it up to the potential of the impulse. An accurate compensation of this current can be achieved by integrating the capacitance charging current (12 in Fig. 37) and, from the voltage so obtained, feeding an opposing current into the current-measuring circuit. The influence of capillary response, which arises from a thin film of liquid penetrating between the mercury thread and the inner wall of the capillary, is less pronounced in pulse polarography than in square-wave polarography, as we have already mentioned. At high sensitivities, however, it causes interference. It can be compensated by an opposing current which is proportional to the amplitude of the impulse, and can be regulated empirically (see Fig. 37). This compensation is imperfect, because there is no linear relationship between capillary response and height of the impulse. The same is true for those components of the current that result from the adsorption or desorption of surface-active substances. Even though such currents have shorter decay times than the electrode reaction current, and cause less interference with the pulse polarograph than the square-wave polarograph, it is advisable to exclude them as far as possible.

Oscillation of the mercury drop can be set up as a result of the sudden change in interfacial tension at the start of the impulse. According to the frequency of this oscillation there will be modulation of the background current, and some current may get through the high-pass filters, thereby falsifying the results. This can be minimized by keeping the size of the drop small, and the amplitude of the impulse less than 0.2 to 0.3 V. This is anyhow ensured in recording current peaks, because the height of the impulse is maintained constant at 50 mV. In recording normal polarograms the reference voltage is so adjusted that the wave to be recorded has a maximum amplitude which is still within the limits given.

The *sensitivity* of pulse polarography for *normal polarograms* reaches $10^{-7} M$ independently of the degree of reversibility of the reaction. A divalent depolarizer at this concentration still gives a wave height of 30 mm with the instrument at its maximum sensitivity, while the background variations do not exceed ± 2 mm. The increase in sensitivity compared with ordinary direct-current polarography derives from the fact that the polarizing voltage is not applied during the total drop life, but only when the surface of the drop has almost reached its maximum, as in oscillographic polarography (page 25). The difference is that the full voltage is not applied during the full life of *one* drop, but only during a fraction of it. On the other hand, it differs from strobe polarography, where, although the *measurement* is indeed made during a fractional part of the drop life, the polarization voltage is applied to the electrode during the total drop-time, with the result that there is already a considerably lower concentration gradient when the measurement commences.

The polarograms are smooth curves without serrations caused by the fall of drops, and have horizontal portions corresponding to the reference current and limiting current, so that the precision with which evaluation can be made, particularly at "high" concentrations (10^{-3} to $10^{-4} M$), is noticeably greater than in ordinary direct-current polarography. The ability to obtain normal or wave-shaped polarograms by the method of pulse polarography is furthermore of importance in that such curves are more suitable than current peaks for continuous analytical control of chemical reactions.

In the second method, which records peak currents, the limit of concentration for a reversible electrode reaction is reached at $10^{-8} M$,

B. Nonstationary Methods

and for an irreversible reaction at $5 \times 10^{-8} M$. The gain in sensitivity compared with square-wave polarography is appreciable for irreversible reactions. In addition, the baseline is much smoother, and much more dilute solutions of conducting salt can be used; a concentration of 0.01 M suffices. This is definitely a critical feature of square-wave polarography, because with large amounts of base electrolyte relatively large amounts of electrolytic and surface-active impurities are introduced. Figure 38 enables a comparison to be made between a pulse

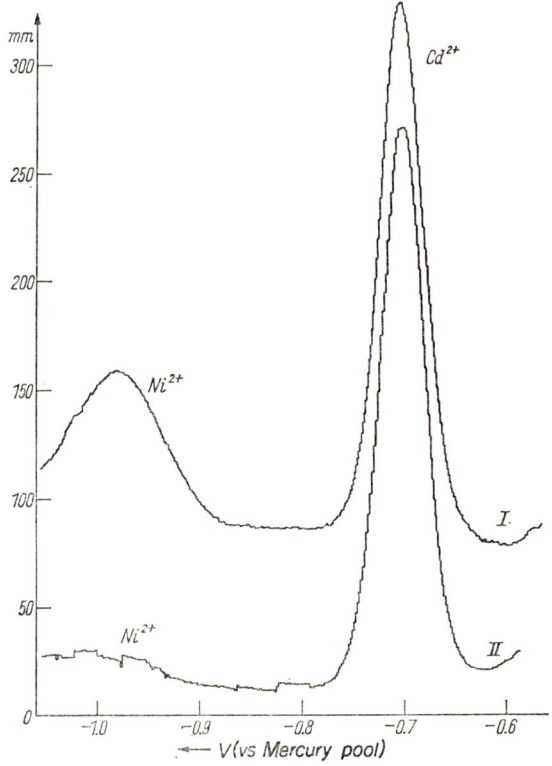

FIG. 38. Pulse polarogram (I) and square-wave polarogram (II) of the same solution ($2 \times 10^{-5} M$ Cd^{2+} and $2 \times 10^{-5} M$ Ni^{2+} in 1 M NH_4Cl/1 M NH_3). Both curves obtained with the maximum sensitivity of the apparatus.

polarogram and a square-wave polarogram. The irreversible nickel wave in the square-wave polarogram is hardly recognizable, while in the pulse polarogram it is quite clear. Another difference is to be seen in the waviness of the reference current.

The *resolvability* is about the same as for alternating-current polarography. The *separability* reaches 1:10⁴, being higher than in conventional alternating-current polarography, and somewhat lower than in square-wave polarography. It is to be noted that this figure applies also to the derivative polarograms. The example in Fig. 39* shows the

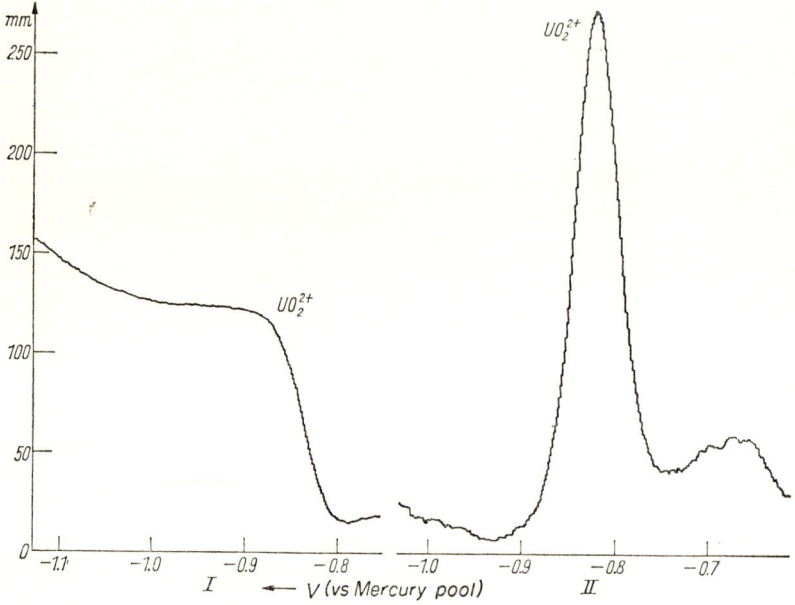

FIG. 39. Wave-pulse polarogram (I) and peak-pulse polarogram (II) of a solution of 10^{-5} M UO_2^{2+} in 4 M H_2SO_4 with an excess of 1.25×10^{-2} M Fe^{3+}. Both polarograms obtained with the maximum sensitivity of the apparatus.

recording of uranyl ions at a concentration of 10^{-5} M in the presence of an excess of Fe^{3+} ions at 1.25×10^{-2} M. As can be seen, there is no difficulty in identifying the UO_2^{2+} either as a peak or as a wave.

Pulse polarography is still comparatively young, and so far detailed reports of experience with the method are not available. Nevertheless, it may be predicted that pulse polarography, in spite of the elaborate nature of the necessary equipment, will establish itself in use, because of its high sensitivity and other valuable features for analytical work.

* We have to thank Dr. G. C. Barker for the information given in Figs. 37, 38 and 39.

II

Methods with Controlled Current

1. Oscillographic Polarography According to Heyrovský and Forejt

It is not easy to find an unambiguous designation for this method, which has been comprehensively described by Heyrovský and Kalvoda.[95] Heyrovský himself calls it "oscillographic polarography with alternating current." If we wish to confine ourselves to prior definitions, it does not belong to oscillographic polarography (Chapter I,B,5), although it makes use of the cathode-ray oscillograph; nor does it belong to alternating-current polarography (Chapter I,B,6), although it operates with alternating current of mains frequency. In fact, it does not really belong to polarography at all, unless one takes as the criterion of a polarographic method the use of a dropping-mercury electrode. Instead of introducing a fresh name, we will retain the one used in the heading to this section, which is generally used in the literature, especially as in this way recognition can be made that the founder of classical polarography has taken an influential part also in the development of polarography in its modern aspects.

The reasons for the development of oscillographic polarography in the true sense have already been given in Chapter I, B, 5. It had the disadvantage compared with conventional polarography that the capacitance current was much greater because of the rapid change in voltage. In the method which they developed, Heyrovský and Forejt,[96] as it were, made a virtue of necessity by controlling the charging and discharging current, and measuring the change in the potential of the electrode.

This procedure is fundamentally different from all those previously considered, in which the voltage is controlled and the current is measured.

Figure 40 shows the principle of the circuit. A sine-wave alternating

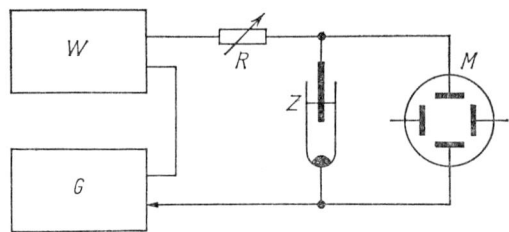

FIG. 40. Schematic diagram for oscillographic polarography (Heyrovský and Forejt).

voltage W of about 100 V, taken from the mains via a transformer, is applied to the cell Z through a high resistor R, which is adjusted so that about 2 V are across the cell. The adjustable direct-current voltage G serves only to fix the *mean* potential of the electrode, so that the potential range lies somewhere between 0 and -2 V with respect to the standard calomel electrode. As the change in resistance of the cell is small compared with the total resistance of the circuit (mainly R), and the changes in potential are small compared with the total voltage (mainly W), the current is in the main decided by the values of W and R; that is to say, it is independent of the changes taking place at the electrode.

On the cathode-ray screen the potential M of the electrode will describe curve I(a) of Fig. 41 during a single period of the change taking place in an "empty" base solution. The lower flat portions of the curve correspond to mercury going into solution at about 0 V; the upper flat portion results from the deposition of conducting salt at about -2 V; the rising and descending portions correspond to the charging and discharging respectively of the electrical double layer. In the absence of the chemical changes associated with the oxidation of mercury and the reduction of the conducting salt the dotted sine curve shown in Fig. 41, I, would be obtained (the same remark applies to Fig. 41, II and III, which will be referred to later). Only the rising

II. Methods with Controlled Current

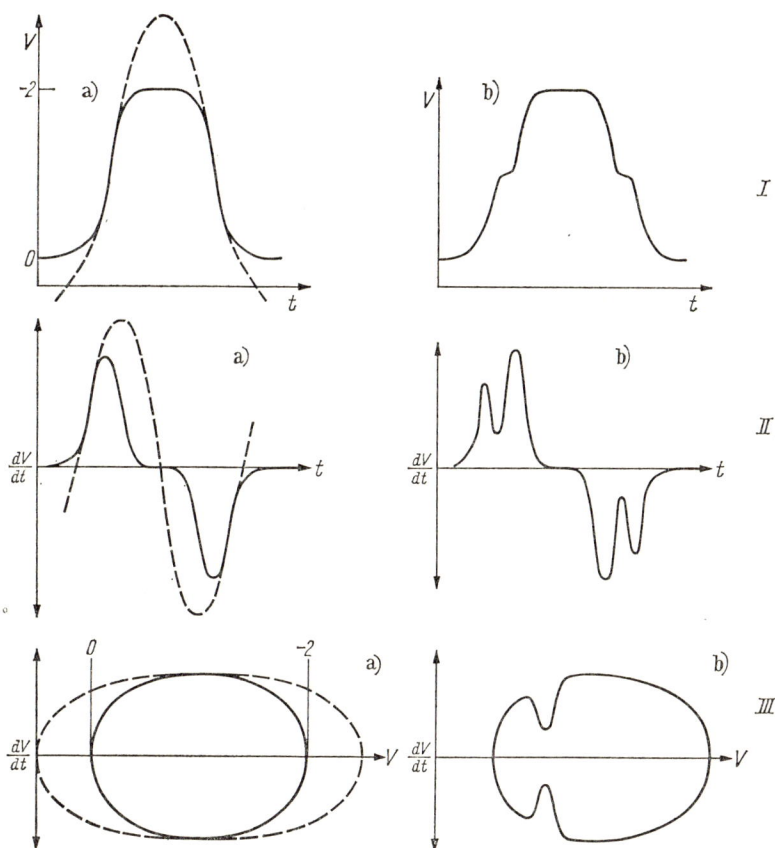

FIG. 41. Various types of oscillographic polarograms. I: $V = f(t)$ curves; II: $dV/dt = f(t)$ curves; III: $dV/dt = f(V)$ curves. Supporting electrolyte with (a), and without (b) reducible depolarizer.

and descending portions of curve I are of interest. The steeper they are, the greater is the value of the controlled current, and the smaller is the capacitance of the electrical double layer. On adding to the solution a substance that is reducible in the polarographic sense, we obtain, if working under the usual polarographic requirements of unpolarizable mercury pool, excess of base electrolyte, etc., a curve of the shape shown in Fig. 41, I(b).

A kink is to be seen on both the rising (cathodic) and descending (anodic) portions of the curve. These rest points start when the deposition potential of the depolarizer in question has been reached, and last until its concentration at the electrode surface has fallen to zero. The voltage then rises again, until another electrode reaction such as the deposition of another depolarizer or the base electrolyte sets in. The production of these rest points can be explained as follows: The total current i is equal to the sum of the charging current i_C and the faradic current i_F; in the absence of depolarizer $i = i_C$. If a depolarizer is reduced, the faradic current flows at the expense of the charging current, that is to say, the potential remains constant, while the charging current i_C and hence dV/dt become zero. The location of the rest points or kinks is characteristic of the type of depolarizer, and their length depends on the concentration of the depolarizer. In contrast with conventional polarography we obtain in this method a kink on the anodic part of the curve corresponding to the oxidation of the previously reduced substance. If both kinks lie at the same level, then the electrode reaction is *reversible*; the greater the degree of *irreversibility*, the greater is the *potential difference* between them. The complete curve therefore provides information on (1) the nature of the depolarizer; (2) its concentration; and (3) the reversibility of the reaction. The quantitative evaluation of these curves is quite inaccurate because the rest points on the very small oscillograms amount to some few millimeters only. Therefore, instead of the simple V-t curves one generally uses their first derivative.

In order to obtain this derivative, the V-t curves are differentiated by means of suitably designed R-C units (see Chapter I,A,3). Figure 41, II, shows the corresponding oscillogram again without (a) and with (b) depolarizer. These curves represent dV/dt vs t.

As the differential capacitance of the electrode is by definition equal to

$$C_D = \frac{dQ}{dV} = \frac{i_C dt}{dV}$$

we can write

$$\frac{dV}{dt} = \frac{i_C}{C_D}$$

that is to say, the function shown is inversely proportional to the differential capacitance. Consequently, even if no faradic current is flowing, dV/dt can change if the differential capacitance alters, as for example by adsorption or desorption of surface-active substances, which occurs at definite potentials. In other words, the method is suitable for the investigation of *adsorption phenomena*. Such adsorption

II. Methods with Controlled Current

and desorption changes are naturally noticeable as correspondingly small kinks on the V-t curves (I), but the effects are much clearer on the dV/dt vs t curves (II). In the case of redox reactions these curves exhibit incisions, which are much better for the purposes of evaluation than the small rest points of the simple V-t curves. On the other hand, the derivative curve is less suitable for qualitative identification of depolarizers, as most of the abscissa is occupied by the uninteresting horizontal portions of the curve. It is reasonable, therefore, to combine both functions V vs t, and dV/dt vs t into a new one, namely, dV/dt vs V, from which the uninteresting component of time is eliminated. According to Heyrovský,[95] it was Ševčík who first had this idea. Figure 41, III, shows the corresponding oscillograms. The abscissa represents the voltage; the ordinate, dV/dt. The dV/dt values in a clockwise sense form a closed curve, again with a cathodic (upper) and an anodic (lower) portion. This method of representation has the additional advantage, with regard to apparatus, that the oscillograph needs no trigger device for the time cycle.

This method has the particular advantage that the equipment needed is modest, and can usually be assembled from available apparatus. Figure 42 shows the arrangement of the "Polaroskop"* which

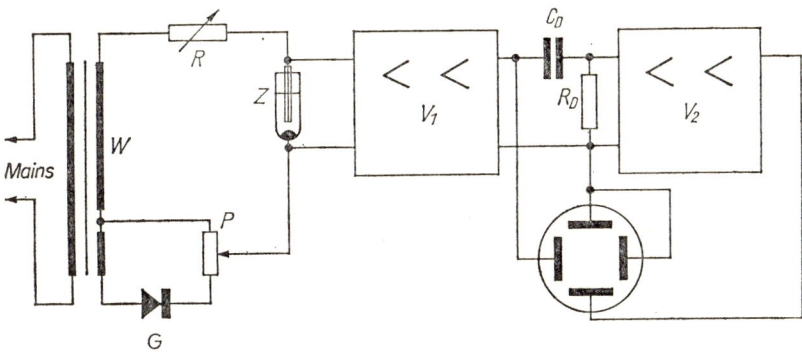

FIG. 42. Schematic diagram of oscillographic polarograph for obtaining $dV/dt = f(V)$ curves.

is manufactured in Prague. A mains transformer supplies the alternating voltage W to the cell Z through the resistor R which has a high resistance; it also supplies the direct-current voltage through the rectifier G and the potentiometer P. The cell voltage is preamplified

* Made by Kovo, Prague.

in the first amplifier V_1, and then applied, on the one hand to the horizontal deflection plates of the cathode-ray tube, and on the other hand to the differentiating unit R_D-C_D. The differentiated voltage after amplification by the second amplifier V_2 is applied to the vertical deflecting plates of the cathode-ray tube. As the voltage contains a range of harmonic frequencies, the time constant of the R-C unit should be at least 1/1000 of the value of the period of the alternating current employed in order to get as near to a true derivative as possible.

A streaming-mercury electrode is frequently employed in oscillographic polarography; in this case the mercury does not fall in drops, but streams upwards at an acute angle through the solution. The container is so constructed that the height of the liquid level and the length of the mercury stream are constant. This electrode has the advantage that it gives static images on the cathode-ray tube in consequence of the constant surface; and they are easier to photograph. The dropping-mercury electrode may be used provided that a sufficiently long drop-time is employed, and exposure takes place shortly before the drop falls. It should be mentioned that in certain circumstances quite different images can be obtained according to whether the dropping electrode or the streaming electrode is used. The reason for this is that reaction products formed at the streaming-mercury electrode are immediately removed, while at the dropping-mercury electrode they linger, and can affect during several periods the current that is sent through the electrode. During this time there can be a series of subsequent reactions, which can produce a very complicated image. On the other hand, the primary electrode process alone is apparent with the streaming-mercury electrode. Finally, in order to avoid polarization phenomena the opposing electrode should not be too small on account of the relatively high current strength (up to several milliamperes).

The method has been considered mathematically several times, by Kambara,[97] Matsuda,[98] and Micka.[99] The theory of Micka agrees best with the experimental results, but only approximately and with a number of limiting assumptions. According to this theory the differential of the electrode voltage is

$$\frac{dE}{dt} = \frac{i_0 \cdot \sqrt{\omega} \cdot \cos \alpha_E}{\frac{n^2 F^2}{4RT}(\sqrt{D_O} c_O + \sqrt{D_R} c_R) + C \cdot \sqrt{\omega}}$$

where i_0 = amplitude of the controlled current; $\omega = 2\pi\nu$ = angular frequency; C = differential capacitance of the electrode; $\alpha_E = \omega t_E - \varphi =$

phase angle at the point of inflection; c = concentration at time $t = 0$. The suffix letters O and R relate to the oxidized and reduced states respectively of the depolarizer.

In deducing this equation it is assumed that the electrolyte does not form complexes with the mercury ions nor with insoluble compounds. Corrections have to be made for these special cases. There is satisfactory agreement between calculated curves and those determined experimentally. Discrepancies between the two can be explained by changes in the capacitance of the electrical double layer, and oscillation of the mercury electrode during the flow of the alternating current. The dependence of the depth of the incision on frequency and temperature to be expected on theoretical grounds can be confirmed approximately by experiment. These matters are of minor importance for analytical purposes, as it is usual to work with calibration curves. Measurement is made of the depth of the incision referred to the potential axis. This depth of the incision is a function of the concentration of the depolarizer, but there is not a linear relationship between the two.

The possible applications of this method are numerous, and far exceed those of conventional polarography. The reasons are as follows:

1. One obtains not only a picture of the *cathodic* electrode changes, but also at the same time one of the *anodic* changes. Apart from being able to draw conclusions regarding the degree of reversibility of the reaction, one also finds it possible to draw far-reaching conclusions regarding the velocity and mechanism of the electrode reaction from the potential separation between the two incisions, from their size, and the dependence of both of these on the composition of the solution, the temperature, and the frequency.

2. The fact that all changes causing an alteration in the differential capacitance likewise produce characteristic incisions extends the range of applicability to numerous substances which themselves are not reducible or oxidizable polarographically, yet in some way are surface-active. We have the same conditions here that were treated in the section on alternating-current polarography under tensammetry. As such adsorbed substances can have a decisive influence on the kinetics of ordinary depolarization changes, in that the degree of reversibility and hence the deposition potential of such reactions can be displaced or the deposition can even be hindered, one has the opportunity of controlling to a great extent the resolvability and separability by a suitable choice of composition of the solution.

3. In consequence of the extraordinarily short time that the products of electrolysis remain at the *streaming*-mercury electrode, the curves that one obtains relate to *primary electrode processes* only, without side reactions or consecutive reactions. It is possible, for instance, to detect the isomers *o*-, *m*-, and *p*-nitrophenol in a mixture (Fig. 43), while in ordinary polarography a single common curve would be obtained.

4. On the other hand, the longer time that the primary reaction products remain at the *dropping*-mercury electrode enables them to react further and lead to fresh incisions, which cannot be observed in classical polarography, where the reverse oxidation change does not occur. Thus, acetylene in alkaline solution gives several characteristic incisions in oscillographic polarography, although polarographically it is not reducible. The incisions are produced because mercury ions formed under a positive potential in an alkaline medium react with acetylene to give acetylides (HgC_2, Hg_2C_2, HgC_2H, etc.), which at more negative potentials are reduced. No incisions are to be seen, naturally, on the anodic portion, because the formation of these acetylides does not take place until mercury goes into solution at a considerable positive potential. The fact that the incisions are produced only at the dropping-mercury electrode shows that one is really dealing with a delayed reaction of this kind. By using a very slowly streaming-mercury electrode the incisions are just visible, but with a rapidly-streaming mercury electrode, where the acetylides after being formed are swept away before they reach their reduction potential, incisions no longer appear. Heyrovský gives the name "artefacts" to these polarographically active substances, which are formed only as a result of electrode reactions at the surface of the electrode.

In order to distinguish among incisions produced by diffusion, kinetic, and adsorption changes, one has recourse to the temperature

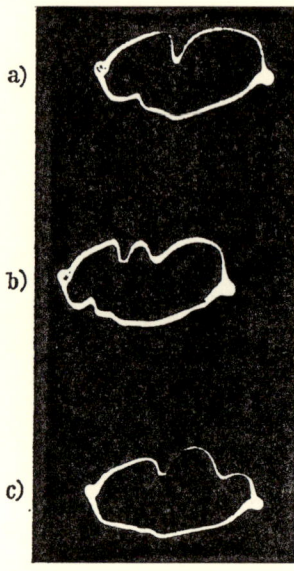

FIG. 43. $dV/dt = f(V)$ curves of $10^{-3}\,M$ nitrophenol in $1\,M$ NaOH at a streaming-mercury electrode. (a) *o*-, (b) *m*-, and (c) *p*-nitrophenol.

II. Methods with Controlled Current

and frequency dependence of the depth of the incision, which has a characteristic form for each of these cases.

The *sensitivity* of oscillographic polarography is about the same as for classical polarography, though it can be considerably increased in special cases if the reduction products form amalgams; the substance to be deposited receives prior electrolytic concentration in a small mercury cathode, and then is immediately subjected to polarographic examination. One measures the time necessary for the disappearance of the incision after having switched from the electrolysis circuit to the polarogram, and determines the concentration by reference to calibration curves that have been made under exactly the same conditions. Concentrations of $10^{-9}\,M$ or lower can be determined with an error of 15–25 %. Kemula and his collaborators[100] in particular have thoroughly investigated this method involving concentration at the hanging-mercury-drop electrode. Naturally, this procedure is not restricted to the method described; Barker and Jenkins,[78] for example, were the first to use it successfully in square-wave polarography.

It is obvious that the *resolvability* and the *separability* are better than in conventional polarography, as (ignoring the simple V-t curves) it is a derivative method. The *accuracy* is naturally not so great as if one took the polarogram with a recorder, because the image is relatively small and the corresponding incision is not so great. The importance of the method described here lies unquestionably in its *qualitative* applications, namely, its ability to identify a multiplicity of reducible and nonreducible substances, and to draw conclusions on reduction, reaction, and adsorption mechanisms. This method, too, is not a substitute for classical polarography; it is a valuable complementary aid.

2. Radio-Frequency Polarography

This method, developed also by Barker,[8] utilizes the so-called faradic rectification, an effect that will be briefly explained. Consider an electrode dipping into a solution in which a reversible redox system is in equilibrium. If we now polarize the electrode by a sine-wave alternating current of small amplitude and radio frequency, we observe a small change in the *mean* potential of the electrode. This change in potential is determined by the asymmetry of the current-voltage curve. If the curve is symmetrical with respect to the equilibrium potential, then the effect does not occur. At least one of the following three inequalities must be satisfied:

$$c_O \neq c_R, \quad D_O \neq D_R, \quad \alpha \neq (1-\alpha)$$

where α = transfer coefficient, D = diffusion coefficient, c = concentration of the components of the reaction in which the suffixes O and R indicate the oxidized and reduced forms. As a comparison one might recall the "reflection" of an alternating-current voltage at the static characteristic line of a valve, which supplies an undistorted alternating current only on the linear portion of this line. Such a simple graphic representation is not possible with the effect in question, because the current-voltage curve is affected by diffusion, and therefore is not "static."

The faradic rectification, also known as the "redox kinetic effect," was first described by Doss and Agarwal,[101] and subsequently thoroughly investigated by Oldham,[102] Hillson,[103] Vdovin,[104] Barker and collaborators,[76,105] and Rangarajan.[106] Matsuda and Delahay[107] treated the inverse case, in which a small alternating-current *voltage* is applied, and the change in the mean *current* is observed.

Mathematical consideration of a redox reaction $O + ne \rightleftharpoons R$ at its equilibrium potential assuming that $V \ll RT/nF$ gives for the value of the displacement of the mean electrode potential

$$\Psi = V^2 \cdot \frac{n \cdot F}{RT} \cdot \left[\frac{2\alpha - 1}{4} + \left(\frac{(1-\alpha)\sqrt{D_O}c_O - \alpha\sqrt{D_R}c_R}{2\sqrt{D_R}c_R + 2\sqrt{D_O}c_O} \right) \cdot \left(\frac{Z \cdot k_e + \sqrt{2Z^2k_e^2}}{\sqrt{2} + 2Zk_e + \sqrt{2Z^2k_e^2}} \right) \right] \quad (1)$$

where

$$Z = \frac{\sqrt{D_R}c_R + \sqrt{D_O}c_O}{c_R\sqrt{\omega}\sqrt{D_R}\sqrt{D_O}},$$

V = amplitude of the alternating component of the electrode potential, $\omega = 2\pi\nu$ = angular frequency of the applied alternating current, and k_e = rate constant of the reduction reaction at the equilibrium potential.

It can be seen immediately from the equation that $\Psi = 0$ when $\alpha = 0.5$, $D_O = D_R$, and $c_O = c_R$; and hence that with a symmetrical current-voltage curve no rectification effect occurs. It is furthermore important that Ψ depends on the values of α and k_e, which are decisive for the kinetics of the electrode reaction, and which can be calculated from it. As the frequency increases, Ψ tends to the limiting value $(2\alpha - 1)V^2nF/4RT$; so that Ψ becomes independent of D and c, and

II. Methods with Controlled Current

α allows itself to be calculated from a knowledge of V. Finally, k_e may be determined from the frequency dependence of Ψ, if the diffusion coefficients are known.

Barker has combined the effect of faradic rectification with the method of square-wave polarography (Chapter I,B,6,b) in the following way. The mean value of the cell voltage is controlled by the square-wave polarograph, as in normal square-wave polarography. The polarization of the mercury electrode is not brought about by a square-wave voltage, but by an amplitude-modulated radio-frequency current of about 200 kc. This radio-frequency current is modulated 100 % by the square-wave voltage of the square-wave polarograph of 225 cycles (see Fig. 45). The displacement of the mean electrode potential in consequence of the faradic rectification is not measured directly, but is probed by the amplitude of the low-frequency alternating current which must be applied to the electrode in order to keep its mean potential constant. This current is amplified and recorded in the usual manner by the square-wave polarograph. Figure 44 shows

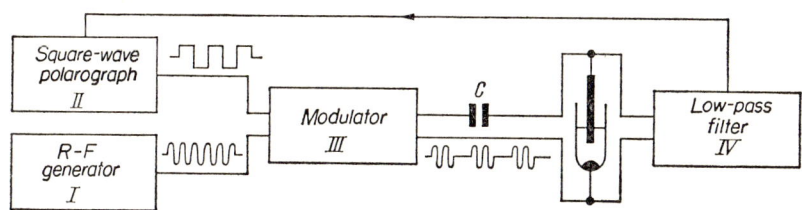

FIG. 44. Schematic diagram for a radio-frequency polarograph.

the arrangement of the circuit. Complete modulation of the radio-frequency voltage generated by the radio-frequency generator I with the square-wave voltage produced by the square-wave polarograph II is effected by the modulator III, and the resulting current is used to polarize the electrode. Once more let it be emphasized that what is measured is not the *change* in the mean *potential*, but the low-frequency alternating *current* resulting from the faradic rectification at *constant* mean potential. This current is fed to the low-pass filter IV of the square-wave polarograph and recorded; the electronic arrangements operate in the same way as previously described. The low-pass filter should prevent any demodulation of the *high* frequency.

The principle of the method may be explained by reference to Fig. 45. The cell usually contains only one of the components O and R of the

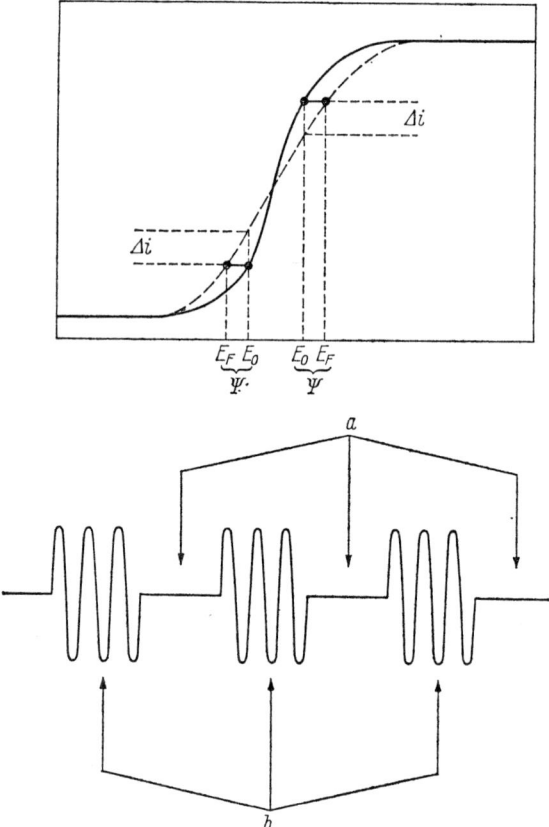

FIG. 45. Diagram illustrating the principle of measurement used in radio-frequency polarography.

reaction; it is generally O. The ratio of concentration of the two at the electrode is varied by a slow change of the electrode potential in the negative direction. The full curve represents the ordinary polarogram; the dotted curve (greatly exaggerated) represents the same polarogram when a radio-frequency current passes through the electrode. At the same mean value of the current the corresponding values of voltage for the two curves differ by the amount Ψ. Let us now polarize the electrode with the modulated radio-frequency current, as shown in the lower part of Fig. 45, and let us suppose that the mean electrode potential is controlled at E_0. During the interval (a) in which no radio-frequency current flows, the potential remains at the value E_0. During

II. Methods with Controlled Current

the interval (b) when the radio frequency is applied, it will attempt to displace itself by an amount Ψ to E_F. It is, however, maintained at E_0 by the square-wave polarograph. To attain this condition, a current Δi must flow; and this current alone is recorded as the polarogram. The measurement of the voltage Ψ is thus reduced to the measurement of a low-frequency alternating current Δi of corresponding amplitude in step with the square-wave voltage. Of course, Ψ may also be measured directly. Two polarograms are then needed, one with and one without radio-frequency polarization; the two are then compared. The method described can however be operated to give a differential curve from which, for each value of electrode voltage or ratio c_O/c_R, a value proportional to Ψ can be obtained. Barker, Faircloth, and Gardner[105] have found that this value corresponds to an alternating current that flows in square-wave polarography when the square-wave polarization voltage has double the amplitude (ΔE), as given by the following expression

$$\Delta E = \frac{(1 - 2\alpha)P^{2\alpha}X^2 + \sqrt{2}P^\alpha(1 - \alpha - \alpha P)X + 1 - P^2}{4[P^{2\alpha}X^2 + \sqrt{2}P^\alpha(1 + P)X + (1 + P)^2]} \cdot \frac{V^2 nF}{RT} \quad (2)$$

where

$$X = \sqrt{\omega \frac{D_O}{K_0}} \quad \text{and} \quad P = \exp{(E - E_{1/2})nF/RT}.$$

K_0 is the rate constant of the reduction reaction at the potential at which $P = 1$. This expression is valid for reactions that are reversible when examined with the square-wave polarograph, and with the condition that only the reaction component O is present in the solution.

The form of the resulting polarograms can vary considerably. Let us assume in the first place that the electrode reaction is controlled by pure diffusion, that is to say, the system is able to follow instantly the rapid changes of the electrode potential. On passing of a radio-frequency current the concentration of the depolarizer will change asymmetrically with respect to the value in the absence of the radio frequency; the mean concentration, and hence the mean reaction velocity, will also be different from the corresponding values without radio-frequency polarization; only at the half-wave potential do they agree. (The example chosen for Fig. 45 relates to this case.) If the radio frequency is now modulated by the square-wave voltage as shown in Fig. 45, practically the same conditions obtain as if a square-wave polarogram were recorded, in which the amplitude of the square-wave voltage changes with the mean electrode potential. The recorded curve will be quite different from a square-wave polarogram, and in the ideal case will have the shape of the second derivative of an ordinary polarogram.

This case hardly arises in practice. In general the reaction will be controlled kinetically by the frequencies employed in radio-frequency polarography, also when it appears reversible in normal polarography. We will now consider the case in which the components of the reaction are not able to follow the rapid changes of the electrode potential. Under these conditions the production of the polarogram shown in Fig. 46[108] may be explained as follows: Curves I and II relate to the

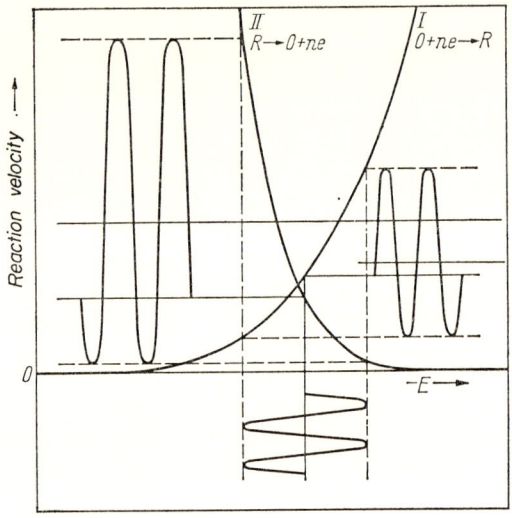

Fig. 46. Diagram illustrating the formation of a radio-frequency polarogram in kinetically controlled electrode reactions (Barker[108]).

dependence of the reaction rates of the forward and reverse reactions (reduction and oxidation) on the voltage. This relationship is given by an exponential function of $\alpha \cdot nF/RT \cdot E$ or $(1 - \alpha) \cdot nF/RT \cdot E$. The example in the figure is valid for $\alpha < 0.5$. A periodic symmetrical change in the potential in consequence of the superimposed radio frequency produces asymmetric periodic changes in the rate of the forward and reverse reactions. Both are increased; the reverse reaction, however, more so, with the result that in all there is a decrease in the current. In the reverse case ($\alpha > 0.5$) there will be an increase; and only when $\alpha = 0.5$ will the current be the same as in the absence of

II. Methods with Controlled Current

radio frequency. While the conditions in the previous "diffusion-controlled" case were such as if there were applied a square-wave voltage whose amplitude varies according to the mean electrode potential, in this, the "kinetic" case, the amplitude is only incidentally dependent on the electrode potential. We now obtain, therefore, as a polarogram a curve which once again for the "ideal case" has the same form as the peak current of an ordinary square-wave polarogram, that is to say, the form of the first derivative of a normal polarogram. Under practical conditions we obtain curves that lie between the two cases described, depending on the frequency and the degree of reversibility of the reaction.

Finally, it must be mentioned that the form of the curve can also be affected by adsorption of one of the components of the reaction, so that under certain conditions, even if $\alpha = 0.5$, a well-defined peak occurs.

Figure 47 shows curves calculated from equations (1) and (2). The

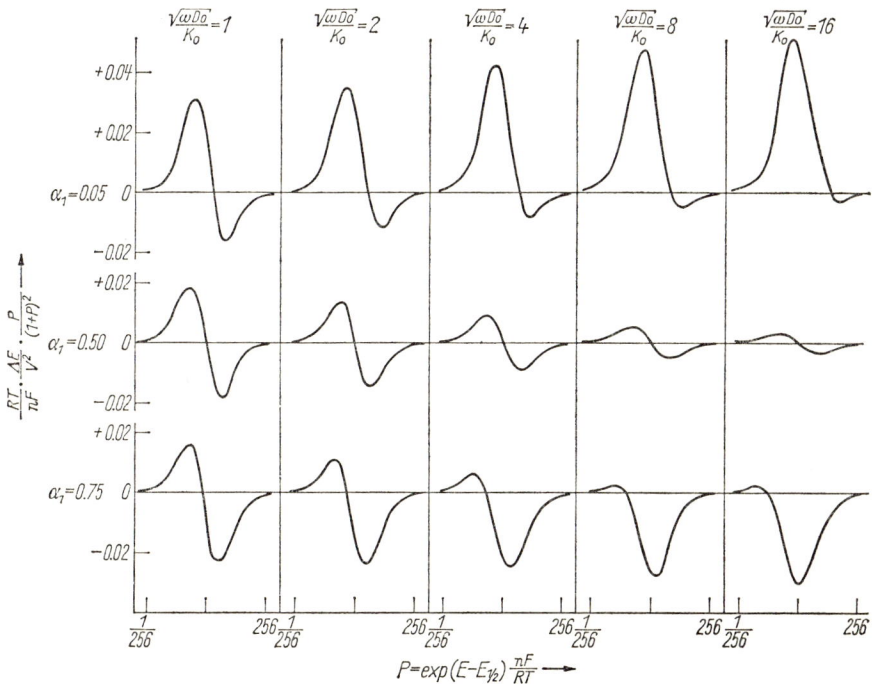

FIG. 47. Influence of frequency and transfer coefficient α on the form of the radio-frequency polarogram (Barker[76]).

abscissa represents $P = \exp(E - E_{1/2}) \cdot nF/RT$, which is, so to speak, the electrode potential standardized with respect to the half-wave potential; the ordinate represents a quantity which is proportional to the measured wave height, namely,

$$\frac{RT}{nF} \cdot \frac{\Delta E}{V^2} \cdot \frac{P}{(1+P)^2}$$

[See Eq. (2) for the meaning of the symbols.] The parameters shown are $(\omega \cdot D_O/K_0)^{1/2}$, a quantity proportional to the frequency (in the horizontal line), and the transfer coefficient α (in the vertical line). Let us consider the middle curves where $\alpha = 0.5$. A "kinetic peak" cannot occur in this case, so we find the diffusion-controlled form of the second derivative, which tends toward zero as the frequency increases, because the concentration changes of the components of the reaction are no longer able to keep step with the rapid changes in potential. When $\alpha < 0.5$ and $\alpha > 0.5$ we still find at lower frequencies a noticeable diffusion effect, which is displaced more and more by the kinetic one as the frequency increases. The curve changes its form from that of the second derivative to that of the first derivative. The value of α determines whether the peak is above or below the axis. As the frequency increases, the peak height in each case tends to a limiting value, which is proportional to $(1 - 2\alpha)$, that is to say, zero when $\alpha = 0.5$.

The *sensitivity* of radio-frequency polarography is greater than that of square-wave polarography, although the recorded current is generally smaller in the former. It has already been mentioned in connection with square-wave polarography that fluctuations in the baseline and hence the effective sensitivity are limited by the capillary-response effect, which is caused by a thin film of liquid entering the capillary and setting up irregular changes in the impedance of the electrode. In radio-frequency polarography this effect operates to much less an extent, so that the ratio signal:baseline and with it the sensitivity are more favorable in spite of the smaller current. The sensitivity for reversible reducible divalent ions is $2 \times 10^{-8}\ M$; for irreversible reducible ions it is $10^{-7}\ M$. The gain in sensitivity in respect of substances that are reduced at highly negative potentials is therefore particularly large, because in such cases the capacitance of the electrical double layer is small, and the radio-frequency component of the electrode potential (V) is large.

Figure 48 gives an example of a square-wave polarogram (above)

II. Methods with Controlled Current

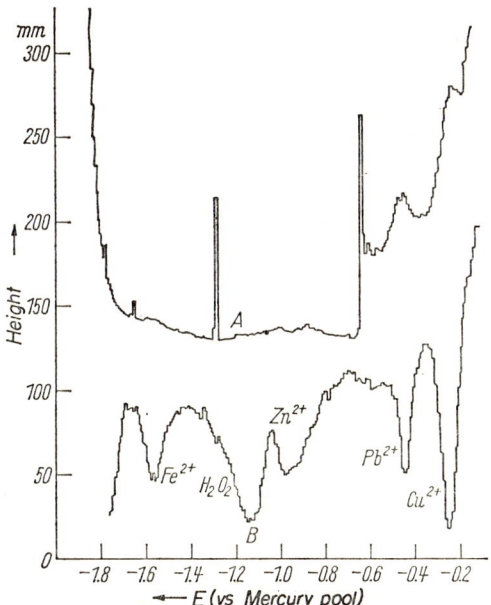

Fig. 48. Square-wave polarogram (A), and radio-frequency polarogram (B) of 1 M KCl. Both at maximum sensitivity. Frequency = 400 cycles. (Barker and Jenkins.[78])

and a radio-frequency polarogram (below) of a solution of 1 M KCl (A.R.). In the square-wave polarogram the Cu and Pb impurities can still just be detected, but the Zn and Fe not at all. In addition the sudden jumps in the baseline resulting from capillary response are clearly seen. On the other hand, in the radio-frequency polarogram all the trace metals are clearly distinguishable against a reference line, which if not smooth is at least continuous.

In order to secure a linear relationship between peak height and concentration the impedance of the boundary layer (at constant potential) must be completely determined by the capacitance of the electrical double layer. Only then is it possible to compute the change in the radio-frequency component (V) of the electrode potential with frequency, and the mean electrode potential (provided that the radio-frequency current density and the relationship between the capacitance of the electrical double layer and the potential are known). A linear relationship between concentration and peak currents is to be expected only at low concentrations of depolarizer, because at high concentrations the capacitance of the electrical double layer is practically independent of the frequency, and only at low concentrations ($<10^{-4}$ M) fully controls the impedance of the boundary layer. The necessity of working at the higher concentrations should hardly arise in practice.

It is only to be expected that radio-frequency polarography should exhibit particularly good *resolvability* and *separability*, because these are advantages of all alternating-current methods in general. Radio-frequency polarography, however, is superior to the other methods, and for the following reasons: We have seen that the height, the form, and the pattern of the recorded currents are prescribed by the kinetic values of the electrode reaction. Different forms of curves (see Fig. 47) are obtained according as the reaction is controlled by diffusion or kinetic processes; and these curves point upward or downward or disappear completely at high frequencies according to the value of α. Substances may very often be identified solely on the basis of the form of their polarograms. For example, Zn^{2+} in a chloride solution gives a "negative" peak, while Ni^{2+} at about the same potential gives a "positive" peak. The kinetic values of the electrode reaction are greatly dependent on the concentration of the solution, so under certain circumstances it is possible to achieve an otherwise very difficult separation of two peaks by a suitable change in the composition of the solution. It can be seen from Fig. 47 that the choice of a suitable frequency likewise plays an important role. Figure 49 provides an example of the excellent resolvability. The separation of indium and cadmium in a chloride solution does not succeed with direct-current polarography; it is quite good in the square-wave polarogram (a); and is much better still in the radio-frequency polarogram (b).

It has been pointed out several times already how important it is, when employing modern polarographic methods, to make a systematic preliminary examination of the problem at hand in order to find the most suitable working conditions. This remark applies with even greater force in radio-frequency polarography.

In conclusion a few words may be said about the possible applications of radio-frequency polarography to kinetic investigations. It has been pointed out that the quantities α and k may be calculated from the relationship existing between frequency and the faradic rectification effect. Barker[105] has made such investigations, and confirmed the close dependence of these quantities on the composition of the solution, a

II. Methods with Controlled Current

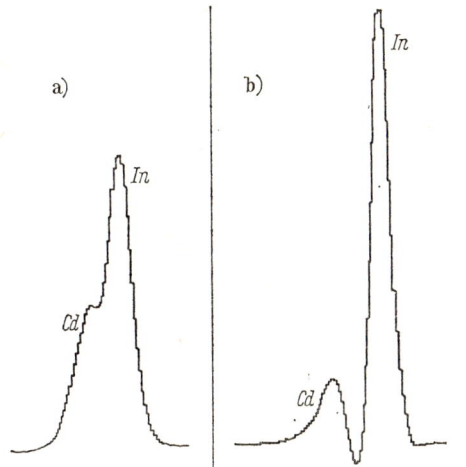

FIG. 49. Square-wave polarogram (A), and radio-frequency polarogram (B) of $4 \times 10^{-5}\,M$ In^{3+} and $8 \times 10^{-5}\,M$ Cd^{2+} in $1\,M$ HCl. Frequency = 400 cycles. (Barker.[8])

fact that can be successfully utilized for the separation analytically of two mutually interfering peak currents, as we have already said. From the results obtained, it is worth noting that many reactions, which are usually considered to take place rapidly, as for example the reduction of Cu^{2+} in $0.5\,M$ sulphate solution, appear to take place relatively slowly by radio-frequency polarography. Insofar as figures are available for the values of kinetic quantities involved in electrode reactions, they are in good agreement with those found by Barker. While the amplitude of the alternating voltage can go up to 25 mV in order to increase the sensitivity for analytical purposes, it should always be kept under 5 mV in kinetic investigations. The frequencies employed lie between 100 kc and 6.4×10^6 cycles. If the frequency is too high another troublesome effect appears; the solution in the proximity of the electrode is heated by the high frequency, with the result that unreproducible irregularities appear in the baseline.

Radio-frequency polarography is only at the beginning of its development; commercially manufactured equipment is not yet available, and published results are still few. It will be the aim of future development to improve the electronic equipment, which is responsible for the fluctuations in the baseline.

A variation that should give promising results would be to use frequency modulation instead of amplitude modulation, whereby the resolvability in particular could be still further increased. In any case it is certain that radio-frequency polarography will be used more and more in the future not only for academic problems but also in the practical field of analysis, notwithstanding the fact that the technique is far from simple: only a layman would have the impression that Fig. 44 involves a simple arrangement of apparatus.

III

Combinations of the Above Methods

The usefulness of the methods previously described can often be increased if they are suitably combined. We have already met some such combinations. For example, square-wave polarography utilizes the advantages of alternating-current polarography, and increases the sensitivity by making use of the same principle (as well as other artifices) as is employed in strobe polarography. It has already been mentioned that electrical differentiation is incorporated in nearly all the methods for the purpose of increasing the resolvability; in oscillographic polarography by the method of Heyrovský the derived curve is the rule.

In all such combinations the advantages and disadvantages of the individual methods remain effective, unless they happen to compensate each other. An example will illustrate this: Mooring[109] has developed a method which combines alternating-current polarography with differential polarography and the single-sweep method of oscillographic polarography. A sine-wave alternating voltage is superimposed on the saw-toothed impulse, and the resulting alternating current, after the direct-current components are filtered out, is applied to the vertical plates of the cathode-ray tube. Moreover, as is usual in differential polarography, two synchronized dropping-mercury electrodes are used in opposition. As differential polarography generally increases the sensitivity, and alternating-current polarography especially increases the resolvability and the separability, it is to be expected that this method would operate to give greater sensitivity and sharper separation than conventional oscillographic polarography. The results do indeed show this to be so. The sensitivity is about 10 times better; and the less noble of two depolarizers can still be determined without difficulty at a concentration ratio of 1:100, while with the ordinary pulse method the separability is quite bad (see, for example, Fig. 13).

It is not necessary to discuss such combinations individually; rather will it suffice to make short reference to them, for the reader himself will be able to form his own idea of the over-all properties from the properties of the individual methods.

Barker and Faircloth[110] have used differential polarography in conjunction with square-wave polarography and radio-frequency polarography. An increase in sensitivity is hardly to be expected, because in these methods a capacitance current is in any case almost without effect. The resolvability and separability are, however, increased; and what is more the reference current is no longer affected to any extent by impurities in the solution of the conducting salt or by traces of oxygen.

Davis and Seaborn[111] have constructed a cathode-ray polarograph using the single-sweep method which can be operated to give differential as well as derivative polarograms. The derivative is obtained by the twin-electrode method, in which the second cell receives a saw-toothed impulse which has a time lag with respect to the first. As the apparatus moreover contains an R-C unit for electrical differentiation, it is possible to record the second derivative of the ordinary polarogram, when both the differentiating units are working. According to Davis and Shalgosky[112] the new equipment is up to 50 times more sensitive than an ordinary cathode-ray polarograph; the resolvability is also much better, and the separability extends even to a concentration ratio of 1000:1.

Barker[8] has also tried to combine the single-sweep method of oscillographic polarography with square-wave polarography with the object of eliminating the capillary response. This effect can vary from drop to drop, so that by using the single-sweep method, which makes the whole polarogram from a *single* drop, no interference can occur. As the magnitude of the capillary response depends not only on the characteristics of the drop at any moment, but also in an unreproducible manner on the potential, the elimination of this undesirable effect is only partial.

Finally, a most interesting apparatus has been developed by Glickstein et al.[113] It combines the method of derivative polarography with a somewhat modified form of strobe polarography. The polariza-

III. Combinations of the Above Methods

tion voltage is applied discontinuously instead of continuously, and by a constant amount (ΔE, for instance, is 5 mV) per drop. The current also is measured discontinuously, and each time shortly before the drop falls; during the remainder of the time the current is stored electronically. The polarogram records the difference (Δi) between two consecutive currents which have been stored in this way and then amplified by a difference amplifier. The relationship $\Delta i/\Delta E$ is obtained directly, and independently of the sweep velocity of the voltage. The method does indeed necessitate synchronization with the drop rhythm, which in this instance is achieved optically by control with a photoelectric cell; but it avoids the disadvantages of other derivative methods, namely the use of two synchronously dropping mercury electrodes, or the distortion of the derivative curve and its exaggerated serrations.

References

1. D. Ilkovič and G. Semerano, *Collection Czechoslov. Chem. Communs.* **4**, 176 (1932).
2. H. Schmidt, *Z. Instrumentenk.* **67**, 301 (1959).
3. P. Delahay, "New Instrumental Methods in Electrochemistry." Interscience, New York, 1954.
4. J. Heyrovský, *Analyst* **81**, 189 (1956).
5. G. Semerano and L. Riccoboni, *Gazz. chim. ital.* **72**, 297 (1942).
6. L. Airey and A. A. Smales, *Analyst* **75**, 287 (1950).
7. M. T. Kelley and H. H. Miller, *Anal. Chem.* **24**, 1895 (1952).
8. G. C. Barker, *Proc. Congr. on Modern Anal. Chem. in Ind., St. Andrews, 1957*, p. 199.
9. J. Heyrovský, *Analyst* **72**, 229 (1947); *Anal. Chim. Acta* **2**, 537 (1948).
10. E. Barendrecht, *Anal. Chim. Acta* **15**, 484 (1956).
11. J. Heyrovský, *Chem. Listy* **43**, 149 (1949).
12. M. P. Lévêque and F. Roth, *J. chim. phys.* **46**, 480 (1949).
13. J. Vogel and J. Říha, *J. chim. phys.* **47**, 5 (1950).
14. J. Říha. *Collection Czechoslov. Chem. Communs.* **16**, 479 (1951).
15. J. J. Lingane and R. Williams, *J. Am. Chem. Soc.* **74**, 790 (1952); consult this for further references.
16. T. Jäckel, *Z. anal. Chem.* **173**, 59 (1960).
17. M. T. Kelley and D. J. Fisher, *Anal. Chem.* **28**, 1130 (1956); see also M. T. Kelley, D. J. Fisher, W. D. Cooke, and H. C. Jones, *Advances in Polarog.* **1**, 158 (1960).
18. E. M. Skobets and N. S. Kavetskii, *Zavodskaya Lab.* **15**, 1299 (1949).
19. S. Wolf, *Chem. Rundschau* **13**, 3 (1960).
20. S. Wolf. *Angew. Chem.* **72**, 449 (1960).
21. J. Paulik and J. Proszt, *Magyar Kém. Folyóirat* **62**, 220 (1956).
22. E. Wåhlin, *Radiometer Polarog.* **1**, 113 (1952); E. Wåhlin and Å. Bresle, *Acta Chem. Scand.* **10**, 935 (1956).
23. G. C. Barker and I. L. Jenkins, *Analyst* **77**, 685 (1952).
24. K. Kronenberger, H. Strehlow, and A. W. Elbel, *Polarog. Ber.* **5**, 62 (1957).
25. H. Strehlow, *Z. Elektrochem.* **55**, 420 (1951).
26. A. W. Elbel, *Z. anal. Chem.* **173**, 70 (1960).
27. Å. Bresle, *Acta Chem. Scand.* **10**, 943, 947, 951 (1956); *LKB-Sci. Tools* **4**, 33 (1957).
28. L. A. Matheson and N. B. Nichols, *Trans. Electrochem. Soc.* **73**, 193 (1938).
29. K. Cruse and W. Heberle, *Z. Elektrochem.* **57**, 579 (1953).
30. J. E. B. Randles, *Trans. Faraday Soc.* **44**, 327, 334 (1948).
31. F. C. Snowden and H. T. Page, *Anal. Chem.* **22**, 969 (1950).
32. P. Delahay, *J. Phys. & Colloid. Chem.* **53**, 1279 (1949).
33. P. Delahay, *J. Phys. & Colloid. Chem.* **54**, 403 (1950).
34. P. Delahay and G. Perkins, *J. Phys. & Colloid. Chem.* **55**, 586 (1951).
35. G. F. Reynolds and H. M. Davis, *Analyst* **78**, 314 (1953).
36. A. Ševčik, *Collection Czechoslov. Chem. Communs.* **13**, 349 (1948).
37. P. Delahay, *J. Phys. & Colloid. Chem.* **54**, 630 (1950).
38. P. Delahay, *J. Am. Chem. Soc.* **75**, 1190 (1953).

39. T. Berzins and P. Delahay, *J. Am. Chem. Soc.* **75**, 555 (1953).
40. J. E. Strassner and P. Delahay, *J. Am. Chem. Soc.* **74**, 6232 (1952).
41. W. Hans and W. Henne, *Naturwissenschaften* **40**, 524 (1953).
42. J. W. Loveland and P. J. Elving, *J. Phys. Chem.* **56**, 250 (1952).
43. R. Bieber and G. Trümpler, *Helv. Chim. Acta* **30**, 971 (1947).
44. D. J. Ferrett, G. W. C. Milner, H. I. Shalgosky, and S. J. Slee, *Analyst* **81**, 506 (1956).
45. P. Favero, *Atti accad. nazl. Lincei, Rend. Classe sci. fis., mat. e nat.* **14**, 433, 520 (1953).
46. F. C. Snowden and H. T. Page, *Anal. Chem.* **24**, 1152 (1952).
47. K. Cruse, *Polarog. Ber.* **3**, 139 (1955).
48. R. H. Müller, R. L. Garman, M. E. Droz, and J. Petras, *Ind. Eng. Chem., Anal. Ed.* **10**, 339 (1938).
49. J. Boeke and H. van Suchtelen, *Philips' tech. Rundschau* **4**, 243 (1939).
50. J. Boeke and H. van Suchtelen, *Z. Elektrochem.* **45**, 753 (1939).
51. C. MacAleavy, Belgian Patent 443,003 (1941); French Patent 886,848 (1942).
52. B. Breyer and F. Gutmann, *Australian J. Sci.* **8**, 21, 163 (1945).
53. B. Breyer and F. Gutmann, *Trans. Faraday Soc.* **42**, 645, 650 (1946); **43**, 785 (1947); *Discussions Faraday Soc.* **1**, 19 (1947).
54. H. Matsuda, *Z. Elektrochem.* **62**, 977 (1958).
55. B. Breyer, H. H. Bauer, and S. Hacobian, *Australian J. Chem.* **8**, 322 (1955).
56. H. H. Bauer and P. J. Elving, *Anal. Chem.* **30**, 334 (1958).
57. B. Breyer, F. Gutmann, and S. Hacobian, *Australian J. Sci. Research* **A3**, 567 (1950).
58. D. C. Grahame, *J. Am. Chem. Soc.* **63**, 1207 (1941); **68**, 301 (1946).
59. P. Delahay and T. J. Adams, *J. Am. Chem. Soc.* **74**, 5740 (1952).
60. H. Gerischer, *Z. Elektrochem.* **58**, 89 (1954); J. Schön, W. Mehl, and H. Gerischer, *ibid.* **59**, 144 (1955).
61. H. Schmidt and M. von Stackelberg, *J. Electroanal. Chem.* **1**, 133 (1959).
62. A. Kalyanasundaram, *Proc. Indian Acad. Sci.* **A33**, 316 (1951).
63. H. H. Bauer, *J. Electroanal. Chem.* **1**, 2 (1959).
64. J. E. B. Randles, *Discussions Faraday Soc.* **1**, 11 (1947).
65. D. C. Grahame, *Chem. Rev.* **41**, 441 (1947).
66. J. van Cakenberghe, *Bull. soc. chim. Belg.* **60**, 3 (1951).
67. H. H. Bauer and P. J. Elving, *Anal. Chem.* **30**, 341 (1958).
68. K. S. G. Doss and S. L. Gupta, *Proc. Indian Acad. Sci.* **A36**, 493 (1952).
69. B. Breyer and S. Hacobian, *Australian J. Sci. Research* **A5**, 500 (1952).
70. B. Breyer and S. Hacobian, *Australian J. Chem.* **9**, 7 (1956).
71. B. Breyer and S. Hacobian, *Anal. Chim. Acta* **16**, 497 (1957).
72. B. Breyer and S. Hacobian, *Australian J. Chem.* **6**, 186 (1953).
73. B. Breyer, H. H. Bauer, and S. Hacobian, *Australian J. Chem.* **7**, 305 (1954).
74. K. Schwabe and H. Jehring, *Z. anal. Chem.* **173**, 36 (1960).
75. F. von Sturm and M. Ressel, *Microchem. J.* **5**, 53 (1961).
76. G. C. Barker, *Anal. Chim. Acta* **18**, 118 (1958).
77. T. Biegler and B. Breyer, *Rev. Polarog.* **7**, 31 (1959).
78. G. C. Barker and I. L. Jenkins, *Analyst* **77**, 685 (1952).
79. R. Kalvoda J. Macků, and K. Micka, *Z. physik. Chem. (Leipzig)* Spec. Issue, p. 66 (1958).

80. M. Kalousek, *Collection Czechoslov. Chem. Communs.* **13**, 322 (1955).
81. T. Kambara, *Bull. Chem. Soc. Japan* **27**, 523 (1954).
82. J. Koutecký, *Collection Czechoslov. Chem. Communs.* **21**, 433 (1953).
83. G. C. Barker, R. L. Faircloth, and A. W. Gardner, Atomic Energy Research Establishment Report C/R 1786. H. M. Stationary Office, London, 1956.
84. J. Koutecký, *Collection Czechoslov. Chem. Communs.* **18**, 597 (1953).
85. D. J. Ferrett, G. W. C. Milner, H. I. Shalgosky, and L. J. Slee, *Analyst* **81**, 506 (1956).
86. F. von Sturm, *Z. anal. Chem.* **166**, 100 (1959); **173**, 11 (1960).
87. M. Kaukewitsch and F. von Sturm, *Advances in Polarog.* **2**, 551 (1960).
88. A. A. Vlček, *Collection Czechoslov. Chem. Communs.* **19**, 862 (1954).
89. F. von Sturm, *J. Polarog. Soc.* **2**, 28 (1958).
90. T. Takahashi and E. Niki, *Talanta* **1**, 245 (1958).
91. T. Takahashi and E. Niki, *Talanta* **1**, 177 (1958).
92. D. M. Miller, *Can. J. Chem.* **34**, 942 (1956).
93. S. Oka, Private communication. Cf. also Shimadzu Seisakusho Ltd., Polarographic Discussion, Meeting Electrochemical Society of Japan, Sendai, September 1958.
94. G. C. Barker and A. W. Gardner, *Z. anal. Chem.* **173**, 79 (1960).
95. J. Heyrovský and R. Kalvoda, "Oszillographische Polarographie mit Wechselstrom." Akademie-Verlag, Berlin, 1960; gives additional references.
96. J. Heyrovský and J. Forejt, *Z. physik. Chem.* **193**, 77 (1943).
97. T. Kambara, *Leybolds polarog. Ber.* **2**, 59 (1954).
98. H. Matsuda, *Z. Elektrochem.* **60**, 617 (1956).
99. K. Micka, *Z. physik. Chem.* (Leipzig) **206**, 345 (1957).
100. W. Kemula and Z. Kublik, *Anal. Chim. Acta* **18**, 104 (1958); also a comprehensive survey in *Advances in Polarog.* **1**, 105 (1960).
101. K. S. G. Doss and H. P. Agarwal, *J. Sci. Ind. Research (India)* **A9**, 280 (1950); *Proc. Indian Acad. Sci.* **A34**, 263 (1951); **A35**, 45 (1952).
102. K. B. Oldham, *Trans. Faraday Soc.* **53**, 80 (1957).
103. P. J. Hillson, *Trans. Faraday Soc.* **50**, 385 (1954).
104. Ju. A. Vdovin, *Doklady Akad. Nauk S.S.S.R.* **120**, 554 (1958).
105. G. C. Barker, R. L. Faircloth, and A. W. Gardner, *Nature* **181**, 247 (1958).
106. S. K. Rangarajan, *J. Electroanal. Chem.* **1**, 396 (1960).
107. H. Matsuda and P. Delahay, *J. Am. Chem. Soc.* **82**, 1547 (1960).
108. G. C. Barker, *Advances in Polarog.* **1**, 144 (1960).
109. C. J. Mooring, *Polarog. Ber.* **6**, 63 (1958).
110. G. C. Barker and R. L. Faircloth, *J. Polarog. Soc.* **2**, 11 (1958).
111. H. M. Davis and J. E. Seaborn, *Advances in Polarog.* **1**, 239 (1960).
112. H. M. Davis and H. I. Shalgosky, *Advances in Polarog.* **2**, 618 (1960).
113. J. Glickstein, S. Rankowitz, C. Auerbach, and H. L. Finston, *Advances in Polarog.* **1**, 183 (1960).
114. H. W. Nürnberg and M. von Stackelberg, *J. Electroanal. Chem.* **2**, 181, 350 (1961); **4**, 1 (1962).
115. H. W. Nürnberg, *J. Anal. Chem.* **186**, 1 (1962).

Author Index

Numbers in parentheses are reference numbers and are inserted to assist in locating a reference when the author's name is not cited in the text. Numbers in italic indicate the page on which the complete reference is listed.

A

Adams, T. J., 37, *94*
Agarwal, H. P., 80, *95*
Airey, L., 7, 11, 23, *93*
Auerbach, C., 92 (113), *96*

B

Barendrecht, E., 9, *93*
Barker, G. C., 7, 14, 47, 51, 54, 55, 56, 57, 63, 79, 80, 83, 84, 85, 87, 88, 89, 92, *93*, *94*, *95*, *96*
Bauer, H. H., 33 (55), 34, 35, 39, 40, 43 (73), 49 (55), *94*, *95*
Berzins, T., 21, *94*
Bieber, R., 24 (43), *94*
Biegler, T., 51, *95*
Boeke, J., 29, *94*
Bresle, Å., 14, 16, *94*
Breyer, B., 30, 33, 35, 36 (57), 39, 40 (57), 42 (70), 43 (71), 49, 50, 51, *94*, *95*

C

Cooke, W. D., 11 (17), *93*
Cruse, K., 19, 21, 22, 29, *94*

D

Davis, H. M., 19 (35), 92, *94*, *96*
Delahay, P., 3, 19 (32, 33, 34), 20, 21, 22, 23, 24 (38), 39, 80, *93*, *94*, *96*
Doss, K. S. G., 42, 80, *95*
Droz, M. E., 29 (48), 30 (48), 40 (48), *94*

E

Elbel, A. W., 14 (24), 16, *94*
Elving, P. J., 24 (42), 34, 35, 40, *94*, *95*

F

Faircloth, R. L., 56 (83), 80 (105), 83, 92, *95*, *96*
Favero, P., 28, *94*
Ferrett, D. J., 20, 27, 57, *94*, *95*
Finston, H. L., 92 (113), *96*
Fisher, D. J., 11, *93*
Forejt, J., 71, 72, *95*

G

Gardner, A. W., 56 (83), 63, 80 (105), 83, *95*, *96*
Garman, R. L., 29 (48), 30 (48), 40 (48), *94*
Gerischer, H., 37, *94*
Glickstein, J., 92, *96*
Grahame, D. C., 36, 39, *94*, *95*
Gupta, S. L., 42, *95*
Gutmann, F., 30, 35 (57), 36 (57), 40 (57), 49 (57), 50 (57), *94*

H

Hacobian, S., 33 (55), 35 (57), 36 (57), 40 (57), 42 (70), 43 (71, 73), 49 (55, 57), 50 (57), *94*, *95*
Hans, W., 23, *94*
Heberle, W., 19, 21, 22, *94*
Henne, W., 23, *94*
Heyrovský, J., 3, 8, 9, 71, 75, *93*, *95*
Hillson, P. J., 80, *95*

I

Ilkovič, D., 1 (1), 52, *93*

J

Jäckel, T., 11, *93*

Jehring, H., 43, *95*
Jenkins, I. L., 14, 51, 79, 87, *94*, *95*
Jones, H. C., 11 (17), *93*

K

Kalousek, M., 54, *95*
Kalvoda, R., 53, 71, *95*
Kalyanasundaram, A., 37, *94*
Kambara, T., 56, 76, *95*
Kaukewitsch, M., 57, *95*
Kavetskii, N. S., 11, *93*
Kelley, M. T., 7, 11, *93*
Kemula, W., 79, *95*
Koutecký, J., 56, 57, *95*
Kronenberger, K., 14, *94*
Kublik, Z., 79 (100), *95*

L

Lévêque, M. P., 9, 13, *93*
Lingane, J. J., 10, *93*
Loveland, J. W., 24 (42), *94*

M

MacAleavy, C., 30, *94*
Macků, J., 53 (79), *95*
Matheson, L. A., 17, 22, *94*
Matsuda, H., 34, 37, 38, 56, 76, 80, *94*, *95*, *96*
Mehl, W., 37 (60), *94*
Micka, K., 53 (79), 76, *95*
Miller, D. M., 59, *95*
Miller, H. H., 7, *93*
Milner, G. W. C., 20 (44), 27 (44), 57 (85), *94*, *95*
Mooring, C. J., 91, *96*
Müller, R. H., 29, 30, 40, *94*

N

Nichols, N. B., 17, 22, *94*
Niki, E., 58, *95*

O

Oka, S., 60, *95*
Oldham, K. B., 80, *95*

P

Page, H. T., 19 (31), 22 (31), 28, *94*
Paulik, J., 11, *93*
Perkins, G., 19 (34), 23, *94*

Petras, J., 29 (48), 30 (48), 40 (48), *94*
Proszt, J., 11, *93*

R

Randles, J. E. B., 19 (30), 20, 39, *94*, *95*
Rangarajan, S. K., 80, *96*
Rankowitz, S., 92 (113), *96*
Ressel, M., 45, 46, 48, 58, *95*
Reynolds, G. F., 19 (35), *94*
Riccoboni, L., 6, *93*
Říha, J., 9, *93*
Roth, F., 9, 13, *93*

S

Schmidt, H., 3, 37, 54, 55, *93*, *94*
Schön, J., 37 (60), *94*
Schwabe, K., 43, *95*
Seaborn, J. E., 92, *96*
Semerano, G., 1 (1), 6, 52, *93*
Ševčik, A., 20, 24, *94*
Shalgosky, H. I., 20 (44), 27 (44), 57 (85), 92, *94*, *95*, *96*
Skobets, E. M., 11, *93*
Slee, S. J., 20 (44), 27 (44), *94*, *95*
Smales, A. A., 7, 11, 23, *93*
Snowden, F. C., 19 (31), 22 (31), 28, *94*
Strassner, J. E., 22, *94*
Strehlow, H., 14 (24), 15, *94*

T

Takahashi, T., 58, *95*
Trümpler, G., 24 (43), *94*

V

van Cakenberghe, J., 40, *95*
van Suchtelen, H., 29, *94*
Vlček, A. A., 57, *96*
Vdovin, Ju. A., 80, *96*
Vogel, J., 9, *93*
von Stackelberg, M., 37, 54, *94*
von Sturm, F., 45, 46, 48, 57, 58, *95*

W

Wåhlin, E., 14, *93*
Williams, R., 10, *93*
Wolf, S., 11, *93*

Subject Index

Adsorption, 32, 41, 74, 85
Alternating current bridge polarography, 58
Alternating current polarography, 29, 91
Amplitude—dependence on peak current, 35, 44
Amalgam formation, 79
Analysis—continuous, 68
Artefacts, 78

Capacitance current, 1, 14, 25, 52, 60, 64, 67, 71
Capacitance—pseudo, 39
Capillary response, 54, 64, 67, 87, 92
Compensation, 6, 7, 16, 25, 54, 60
Concentration—related factors, 11, 12, 20, 38, 44
Curve follower, 7

Depletion effect, 23
Derivative polarography, 8, 27, 74, 92
Desorption—see adsorption
Differential capacitance, 42, 74
Differential polarography, 6, 91, 92
Displacement currents, 32, 42

Electrical double layer—capacitance of, 1, 45, 47, 87

Faradic rectification, 79
Faradic impedance, 36
Frequency—dependence on peak current, 39, 44

Impedance, 36, 39, 44

Kinetics, 16, 27, 39, 77, 84, 88

Multi-sweep method, 22

Oscillographic polarography, 16, 71, 91, 92

Peak current, 10, 20, 31, 34, 37, 44, 66, 84, 85

Peak potential, 20, 31
Polarization resistance, 39
Ψ_1—potential, 47
Pulse polarography, 62

Radio-frequency polarography, 79, 92
Rate constant—computation of, from peak current, 39
Rate constant—computation of, by faradic rectification, 80
Rectification—phase dependent, 36
Redox kinetic effect, 80
Reproducibility, 12, 16, 21
Resistance—dependence on peak current, 36
Resolvability—defined, 2
Reversibility, 10, 21, 33, 38, 39, 57, 74, 77, 84

Sensitivity—defined, 1
Separability—defined, 2
Single-sweep method, 17, 91, 92
Solutions—composition of, 40, 45, 51, 77, 88
Square wave polarography, 51, 92
Square wave voltage, 51
Streaming-mercury electrode, 7, 76
Strobe polarography, 13, 52, 92
Surface-active substances, 43, 50, 67, 77
Synchronization, 7, 14, 18, 22, 56, 67, 93

"Tastpolarographie," 13, 52, 92
Tensammetry, 42
Transfer coefficient—determination of, using harmonics, 40
Transfer coefficient—computation of, by faradic rectification, 80, 88
Triangular voltage, 24